ECONOMIC ANALYSIS
AND
INVESTMENT DECISIONS

ECONOMIC ANALYSIS
AND
INVESTMENT DECISIONS

Chi U. Ikoku
The University of Port Harcourt

John Wiley & Sons

New York   *Chichester*   *Brisbane*   *Toronto*   *Singapore*

Library of Congress Cataloging in Publication Data:

Ikoku, Chi U.
  Economic analysis and investment decisions.

  Includes bibliographical references and index.
   1. Petroleum industry and trade—Finance.   2. Gas. industry—
Finance.   3. Investments.   I. Title.
HD9560.5.138   1985   622′.338′0681   84-26980
ISBN 0-471-81455-5

Printed in the United States of America

10 9 8 7 6 5 4 3 2 1

To the memory of our beloved son
Uchenna Chukwuemeka Ikoku

PREFACE

The aim of this book is to present the techniques involved in the economic evaluation of drilling prospects and oil and gas producing properties. It is intended to serve as a basic textbook on the subject of petroleum economics, one that will be used by undergraduate students in petroleum engineering and petroleum geology programs. It can also be used as a reference work or self-study textbook by practicing engineers, geologists, and managers.

This book presents practical methods, reinforced by numerous examples for analyzing decisions concerning investments in the oil and gas industry. Chapter 1 is an introduction to economic analysis and decision making and sets the stage for the rest of the book. Chapter 2 discusses volumetric estimates, material balance estimates, and empirical correlations for oil and gas reserves in place. Chapter 3 treats the important subject of production forecasting by decline curve analysis.

Some terms and concepts related to the financial aspects of a business are introduced in Chapter 4. The procedures for making accurate cash flow projections are also discussed in some detail in this chapter. The concept of time value of money is introduced in Chapter 5. This chapter summarizes compound interest relationships and discount factor tables that are needed in economic evaluation of investment proposals.

Making sound capital expenditure decisions requires an objective means of measuring the productivity of individual investment proposals. Chapter 6 discusses the common profit indicators, their advantages and weaknesses. Applications of these measures of profitability are given in detail in Chapter 7.

Chapter 8 is concerned with risk and uncertainty. Several techniques of quantifying risk including mathematical expectation, decision trees, utility theory, and simulation are discussed. This chapter discusses probability distributions and how they can be used to represent uncertain data.

The material on which this book is based has been used successfully in petroleum and natural gas engineering school and adult education courses in the United States and overseas. I am indebted to my colleagues and also to my students whose enthusiasm for the subject has made teaching a pleasure. Their constructive criticisms and comments became textbook inputs. Special thanks must also be given to Mrs. Peggy Conrad for typing, correcting, sometimes editing, always contributing to the final manuscript.

I would like to express my appreciation to the editorial staff of John Wiley, including Merrill Floyd and Bruce Safford, for their patience and politeness. I thank Ellen Baron and members of Wiley's production staff for a fine job.

Chi U. Ikoku

CONTENTS

NOMENCLATURE

QUANTITIES IN ALPHABETICAL ORDER

(*) Dimensions: L = length, m = mass, q = electrical charge, t = time, and T = temperature.

(**) To avoid conflicting designation in some cases, use of reserve symbols and reserve subscripts is permitted.

Quantity	SPE Standard	Reserve SPE Letter Symbols**	Dimensions*
air requirement	a	F_a	
angle	α alpha	β beta	
angle	θ theta	γ gamma	
angle, contact	θ_c theta	γ_c gamma	
angle of dip	α_d alpha	θ_d theta	
area	A	S	L^2
Arrhenius reaction rate velocity constant	w	z	L^3/m
breadth, width, or (primarily in fracturing) thickness	b	w	L
burning-zone advance rate	v_b	V_b, u_b	L/t
capillary pressure	P_c	P_c, p_c	m/Lt^2
charge	Q	q	q
coefficient, convective heat transfer	h	h_h, h_T	m/t^3T
coefficient, heat transfer, interphase convective (use h, or convective coefficient symbol, with pertinent phase subscripts added)			m/t^3T
coefficient, heat transfer, over-all	U	U_T, U_θ	m/t^3T
coefficient, heat transfer, radiation	I	I_T, I_θ	m/t^3T
components, number of	C	n_c	
compressibility	c	k, κ kappa	Lt^2/m
compressibility factor	z	Z	
concentration	C	c, n	various

Courtesy of Society of Petroleum Engineers of American Institute of Mining, Metallurgical, and Petroleum Engineers, Inc.

Quantity	SPE Standard	Reserve SPE Letter Symbols**	Dimensions*
condensate or natural gas liquids content	C_L	c_L, n_L	various
conductivity	σ sigma	γ gamma	various
conductivity, thermal (always with additional phase or system subscripts)	k_h	λ lambda	mL/t^3T
contact angle	θ_c theta	γ_c gamma	
damage ratio ("skin" conditions relative to formation conditions unaffected by well operations)	F_s	F_d	
density	ρ rho	D	m/L^3
depth	D	y, H	L
diameter	d	D	L
diffusion coefficient	D	μ mu, δ delta	L^2/t
dimensionless fluid influx function, linear aquifer	Q_{LtD}	Q_{ltD}	
dispersion coefficient	K	d	L^2/t
displacement	s	L	L
displacement ratio	δ delta	F_d	
distance between adjacent rows of injection and production wells	d	L_d, L_2	L
distance between like wells (injection or production) in a row	a	$L_a L_1$	L
distance, length, or length of path	L	s, l script l	L
efficiency	E	η eta, e	
electrical resistivity	ρ rho	R	mL^3/tq^2
electromotive force (voltage)	E	V	mL^2/t^2q
elevation referred to datum	Z	D, h	L
encroachment or influx rate	e	i	L^3/t
energy	E	U	mL^2/t^2
enthalpy (always with phase or system subscripts)	H	I	mL^2/t^2
enthalpy (net) of steam or enthalpy above reservoir temperature	H_s	I_s	mL^2/t^2
enthalpy, specific	h	i	L^2/t^2
entropy, specific	s	σ sigma	L^2/t^2T
entropy, total	S	σ_t sigma	mL^2/t^2T
equilibrium ratio	K	k, F_{eq}	
fluid influx function, linear aquifer, dimensionless	Q_{LtD}	Q_{ltD}	
flow rate or flux, per unit area (volumetric velocity)	u	ψ psi	L/t
flow rate or production rate	q	Q	L^3t
fluid (generalized)	F	f	various
flux	u	ψ psi	various
force	F	Q	mL/t^2
formation volume factor	B	F	

Quantity	SPE Standard	Reserve SPE Letter Symbols**	Dimen-sions*
fraction gas	f_g	F_g	
fraction liquid	f_L	F_L, f_l	
frequency	f	ν nu	l/t
fuel consumption	m	F_F	various
fuel deposition rate	N_R	N_F	m/L^3t
gas (any gas, including air)—always with iden-tifying subscripts	G	g	various
gas in place in reservoir, total initial	G	g	L^3
gas−oil ratio, producing (if needed, the reserve symbols could be applied to other gas−oil ratios)	R	F_g, F_{go}	
general and individual bed thickness	h	d, e	L
gradient	g	γ gamma	various
heat flow rate	Q	q, Φ phi$_{cap}$	mL^2/t^3
heat of vaporization, latent	L_v	λ_v lambda	L^2/t^2
heat or thermal diffusivity	α alpha	α, η_h eta	L^2/t
heat transfer coefficient, convective	h	h_h, h_T	m/t^3T
heat transfer coefficient, interphase convec-tion (use h, or convective coefficient symbol with pertinent subscripts added)			m/t^3T
heat transfer coefficient, overall	U	U_T, U_θ	m/t^3T
heat transfer coefficient, radiation	I	I_T, I_θ	m/t^3T
height (elevation)	Z	D, h	L
height (other than elevation)	h	d, e	L
hydraulic radius	r_H	R_H	L
index of refraction	n	μ mu	
influx (encroachment) rate	e	i	L^3/t
influx function, fluid, linear aquifer, dimen-sionless	Q_{LtD}	Q_{ltD}	
initial water saturation	S_{wi}	ρ_{wi} rho, S_{wi}	
injectivity index	I	i	L^4t/m
intercept	b	Y	various
interfacial or surface tension	σ sigma	y, γ gamma	m/t^2
interstitial-water saturation in oil band	S_{wo}	S_{wb}	
irreducible water saturation	S_{iw}	ρ_{iw} rho, S_{iw}	
kinematic viscosity	ν nu	N	L^2/t
length	L	s, l script l	L
length, path length, or distance	L	s, l script l	L
mass flow rate	w	m	m/t
mobility ratio	M	F_λ	
mobility ratio, diffuse-front approximation, $[(\lambda_D + \lambda_d)_{swept}/(\lambda_d)_{unswept}]$; D signifies displacing; d signifies displaced; mobilities are evaluated at average saturation condi-tions behind and ahead of front	$M_{\bar{S}}$	M_{Dd}, M_{su}	

Quantity	SPE Standard	Reserve SPE Letter Symbols**	Dimensions*
mobility ratio, sharp-front approximation, (λ_D/λ_d)	M	F_λ	
mobility ratio, total, $[(\lambda_t)_{swept}/(\lambda_t)_{unswept}]$; "swept" and "unswept" refer to invaded and uninvaded regions behind and ahead of leading edge of a displacement front	M_t	$F_{\lambda t}$	
mobility, total, of all fluids in a particular region of the reservoir; e.g., $(\lambda_o + \lambda_g + \lambda_w)$	λ_t lambda	Λ lambda$_{cap}$	$L^3 t/m$
modulus, bulk	K	K_b	m/Lt^2
modulus of elasticity in shear	G	E_s	m/Lt^2
modulus of elasticity (Young's modulus)	E	Y	m/Lt^2
mole fraction gas	f_g	F_g	
mole fraction liquid	f_L	F_L, f_l	
molecular refraction	R	N	L^3
moles, number of	n	N	
moles of liquid phase	L	n_L	
moles of vapor phase	V	n_v	
moles, total	n	n_t, N_t	
number (of moles, or components, or wells, etc.)	n	N	
oil (always with identifying subscripts)	n	n	various
oil in place in reservoir, initial	N	n	L^3
oxygen utilization	e_{o_2}	E_{o_2}	
path length, length, or distance	L	s, l script l	L
permeability	k	K	L^2
Poisson's ratio	μ mu	ν nu, σ sigma	
porosity	ϕ phi	f, ε epsilon	
pressure	p	P	m/Lt^2
production rate or flow rate	q	Q	L^3/t
productivity index	J	j	$L^4 t/m$
quality (usually of steam)	f_s	Q, x	
radial distance	Δr	ΔR	L
radius	r	R	L
radius, hydraulic	r_H	R_H	L
ratio, damage ("skin" conditions relative to formation conditions unaffected by well operations)	F_s	F_d	
ratio initial reservoir free gas volume to initial reservoir oil volume	m	F_{Fo}, F_{go}	
ratio, mobility	M	F_λ	
ratio, mobility, diffuse-front approximation, $[(\lambda_D + \lambda_d)_{swept}/(\lambda_d)_{unswept}]$; D signifies displacing; d signifies displaced; mobilities are evaluated at average saturation conditions behind and ahead of front	$M_{\bar{s}}$	M_{Dd}, M_{su}	

Quantity	SPE Standard	Reserve SPE Letter Symbols**	Dimensions*
ratio, mobility, sharp-front approximation, (λ_D/λ_d)	M	$F\lambda$	
ratio, mobility, total, $[(\lambda_t)_{swept}/(\lambda_t)_{unswept}]$; "swept" and "unswept" refer to invaded and uninvaded regions behind and ahead of leading edge of a displacement front	M_t	$F_{\lambda t}$	
reaction rate constant	k	r, j	L/t
reciprocal formation volume factor, volume at standard conditions divided by volume at reservoir conditions	b	f, F	
reciprocal permeability	j	ω omega	$1/L^2$
resistance	r	R	mL^2/tq^2
resistance	r	R	various
resistivity, electrical	ρ rho	R	mL^3/tq^2
saturation	S	ρ rho, s	
saturation, water, initial	S_{wi}	$ρ_{wi}$ rho, s_{wi}	
saturation, water, irreducible	S_{iw}	$ρ_{iw}$ rho, s_{iw}	
skin effect	s	S, σ sigma	
skin (radius of well damage or stimulation)	r_s	R_s	L
slope	m	A	various
specific gravity	γ gamma	s, F_s	
specific heat (always with phase or system subscripts)	C	c	L^2/t^2T
specific heats ratio	γ gamma	k	
specific injectivity index	I_s	i_s	L^3t/m
specific productivity index	J_s	j_s	L^3t/m
specific volume	v	v_s	L^3/m
specific weight	F_{wv}	γ gamma	m/L^2t^2
stimulation radius of well (skin)	r_s	R_s	L
strain, normal and general	ε epsilon	e, ϵ_n epsilon	
strain, shear	γ gamma	ϵ_s epsilon	
strain, volume	θ theta	$θ_v$ theta	
stress, normal and general	σ sigma	s	m/Lt^2
stress, shear	τ tau	s_s	m/Lt^2
surface tension	σ sigma	$y, γ$ gamma	m/t^2
temperature	T	θ theta	T
thermal conductivity (always with additional phase or system subscripts)	k_h	λ lambda	mL/t^3T
thermal cubic expansion coefficient	β beta	b	$1/T$
thermal or heat diffusivity	α alpha	$a, η_b$ eta	L^2/t
thickness (general and individual bed)	h	d, e	L
time	t	τ tau	t
total mobility of all fluids in a particular region of the reservoir; e.g., $(\lambda_o + \lambda_g + \lambda_w)$	λ_t lambda	Λ lambda$_{cap}$	L^3t/m

Quantity	SPE Standard	Reserve SPE Letter Symbols**	Dimensions*
total mobility ratio, $[(\lambda_t)_{swept}/(\lambda_t)_{unswept}]$; "swept" and "unswept" refer to invaded and uninvaded regions behind and ahead of leading edge of a displacement front	M_t	$F_{\lambda t}$	
transfer coefficient, convective heat	h	h_h, h_T	m/t^3T
transfer coefficient, heat, interphase convective (use h, or convective coefficient symbol with pertinent phase subscripts added)			m/t^3T
transfer coefficient, heat, overall	U	U_T, U_θ	m/t^3T
transfer coefficient, heat, radiation	I	I_T, I_θ	m/t^3T
utilization, oxygen	e_{O_2}	E_{O_2}	
velocity	v	V, u	L/t
viscosity	μ mu	η eta	m/Lt
volume	V	v	L^3
volumetric velocity (flow rate or flux, per unit area)	u	ψ psi	L/t
water (always with identifying subscripts)	W	w	various
water in place in reservoir, initial	W	w	L^3
water saturation, initial	S_{wi}	ρ_{wi} rho, s_{wi}	
water saturation, irreducible	S_{iw}	ρ_{iw} rho, s_{iw}	
wave number	σ sigma	$\bar{v}$	$1/L$
weight	W	w, G	mL/t^2
wet-gas content	C_{wg}	c_{wg}, n_{wg}	various
width, breadth, or (primarily in fracturing) thickness	b	w	L
work	W	w	mL^2/t^2

Subscripts

Subscript	SPE Standard	Reserve SPE Letter Subscripts**
air	a	A
atmospheric	a	A
average of mean saturation	$\bar{S}$	$\bar{\rho}$ rho, $\bar{s}$
band or oil band	b	B
base	b	r, β beta
boundary conditions, external	e	o
breakthrough	BT	bt
bubble point or saturation	b	s
burned or burning	b	B
calculated	C	calc

Subscript	SPE Standard	Reserve SPE Letter Subscripts**
capillary (usually with capillary pressure, P_c)	c	C
casing or casinghead	c	cg
contact (usually with contact angle, θ_c)	c	C
core	c	C
cumulative influx (encroachment)	e	i
damage or damaged (includes "skin" conditions)	s	d
depleted region, depletion	d	δ delta
dispersed	d	D
dispersion	K	d
displaced	d	s, D
displacing or displacement	D	s, σ sigma
entry	e	E
equivalent	eq	EV
estimated	E	est
experimental	E	EX
fill-up	F	f
finger or fingering	f	F
flash separation	f	F
fraction or fractional	f	r
fracture, fractured, or fracturing	f	F
free (usually with gas or gas-oil ratio quantities)	F	f
front, front region, or interface	f	F
gas	g	G
gross	t	T
heat or thermal	h	T, θ theta
hole	h	H
horizontal	H	h
hydrocarbon	h	H
imbibition	I	i script i
influx (encroachment), cumulative	e	i
injected, cumulative	i	I
injection, injected, or injecting	i	inj
inner or interior	i	ι iota, i script i
interface, front region, or front	f	F
interference	I	i, i script i
invaded	i	I
invaded zone	i	I
invasion	I	i
irreducible	i	i script i, ι iota
linear, lineal	L	l script l
liquid or liquid phase	L	l script l
lower	l script l	L
mean or average saturation	$\bar{S}$	$\bar{\rho}$ rho, $\bar{s}$
mixture	M	m
mobility	λ lambda	M

Subscript	SPE Standard	Reserve SPE Letter Subscripts**
nonwetting	*nw*	*NW*
normalized (fractional or relative)	*n*	*r, R*
oil	*o*	*n*
outer or exterior	*e*	*o*
permeability	*k*	*K*
pore (usually with volume, V_p)	*p*	*P*
production period (usually with time, t_p)	*p*	*P*
radius, radial, or radial distance	*r*	*R*
reference	*r*	*b*, ρ rho
relative	*r*	*R*
reservoir	*R*	*r*
residual	*r*	*R*
saturation, mean or average	$\bar{S}$	$\bar{\rho}$ rho, $\bar{s}$
saturation or bubble point	*b*	*s*
segregation (usually with segregation rate, q_s)	*s*	*S*, σ sigma
shear	*s*	τ tau
skin (stimulation or damage)	*s*	*S*
slip or slippage	*s*	σ sigma
solid(s)	*s*	σ sigma
stabilization (usually with time)	*s*	*S*
steam or steam zone	*s*	*S*
stimulation (includes "skin" conditions)	*s*	*S*
storage or storage capacity	*S*	*S*, σ sigma
strain	ε epsilon	*e*
surface	*s*	σ sigma
swept or swept region	*s*	*S*, σ sigma
system	*s*	σ sigma
temperature	*T*	*h*, θ theta
thermal (heat)	*h*	*T*, θ theta
total, total system	*t*	*T*
transmissibility	*T*	*t*
treatment or treating	*t*	τ tau
tubing or tubing head	*t*	*tg*
unswept or unswept region	*u*	*U*
upper	*u*	*U*
vaporization, vapor, or vapor phase	*v*	*V*
velocity	*v*	*V*
vertical	*V*	*v*
volumetric or volume	*V*	*v*
water	*w*	*W*
weight	*W*	*w*
wellhead	*wh*	*th*
wetting	*w*	*W*

ECONOMIC ANALYSIS
AND
INVESTMENT DECISIONS

1

ECONOMIC ANALYSIS AND DECISION MAKING

1.1   INTRODUCTION

This book is concerned primarily with decision methods for economic evaluation of alternative investment opportunities in the oil and gas industry. Normally an investment analysis should involve economic analysis, financial analysis, and intangible analysis.

Many times an investment alternative that looks attractive economically may be rejected for financial reasons because funds are not available and cannot be obtained at reasonable interest rates to finance the investment proposal. In addition, intangible considerations such as potential loss of public goodwill, pollution problems, or possible future litigation may cause economically sound investments to be rejected. The emphasis in this book, however, is on economic analysis factors. This is not meant in any way to downgrade the importance of financial and intangible considerations in investment decision making.

The primary objective of oil and gas field operations is not merely to produce oil and gas but rather to make a profit—the maximum profit. Engineers, explorationists, and managers are often so burdened with the routine details of getting the work done, maintaining and operating the facilities that they have little opportunity to study costs and profits and to compare, on a profit basis, the various methods and equipment that may be used in achieving a required objective.

We produce oil and gas by drilling wells, but operating companies do not rush out into the field and just drill wells with whatever equipment may be at hand. They must first consider the chances of finding profitable accumulation of hydrocarbons. They then study and plan the development of the field or property as a whole. What will the average well cost to drill and operate? What are its chances of being a producer? What will be the value of its production? How many wells shall the company drill, that is, how closely spaced should the wells be for

1

maximum profit? For each well drilled the company must first decide whether it will pay out or return its cost plus a reasonable profit. In considering the well cost, what method of drilling, what casing program, and what type of equipment will enable us to complete the well satisfactorily at minimum cost? What production method will achieve minimum unit production cost? These are but a few of the typical problems that confront the petroleum engineer and geologist. The general plan and timing of operations is contingent upon the solution of many such problems.

1.2   THE NEED FOR ECONOMIC ANALYSIS

How important are economic judgment and planning in petroleum exploration and production today? During the post World War II development drilling boom, oil companies were producing some 20 percent net returns on net shareholder investments. Money was available at 3 and 4 percent interest rates, and management perhaps did not feel a need for highly refined profitability criteria for investment decisions. However, in the late 1950s profit margins began to shrink, and drilling and development prospects were harder to find and less profitable to produce. As a direct consequence, there arose a need for better profitability criteria.

It is becoming more obvious that only those companies who exercise sound economic judgment, who use sophisticated evaluation techniques, and who prepare and execute carefully conceived, imaginative plans can expect to survive in petroleum exploration and production in the near future. Based on past figures, the amount of oil and gas that must be discovered to maintain our present standard of living is staggering. This must be accomplished in the face of:

1. Decreasing chance of major discoveries as older areas are drilled up.

2. Increasing finding costs.

3. Impact of inflation.

4. Uncertainties in crude oil and natural gas prices.

5. Government control and regulations.

These factors cause conflicting pressures on the explorationist, engineer, and management, and impede the effort to find new reserves (Figure 1.1).

A need for economic evaluation may arise for various reasons. In most cases, such an evaluation will be made prior to:

1. Actual engineering design of the development of oil and gas fields.

2. Acquisition of a field.

3. Planned change in development or production methods that consequently may affect the production rate and ultimate recovery.

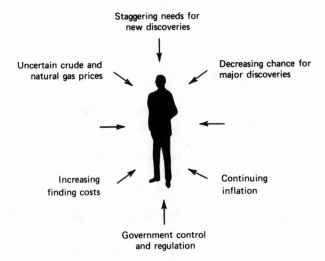

Figure 1.1   Pressures on explorationists.

4.   Assessment of value of assets for taxation purposes.

5.   Reevaluation of priorities in allocation of investment funds by the company.

Whatever the reason for conducting an economic evaluation, with limited financial resources and a specific set of corporate objectives, any company must select the best investment opportunities from among those available. The economic analysis should lead toward an unbiased answer to two questions: (1) Does the particular investment project seem to satisfy the stated objectives of the firm? (2) Is this project "better" or "worse" than other possible projects?

Even if there were only a single investment opportunity under review, it must favorably compare, for optimal allocation of funds, with other profit-generating activities. This concept of *opportunity cost*—the advantage forgone due to alternative use of investment funds—must be considered as an integral part of all economic analyses.

1.3   OBJECTIVES OF THE FIRM

The stated objectives of an oil and gas company considerably influence the selection of capital investment projects. A partial listing of some of the goals of the firm would include:

1.   Maximization of total profits (or minimization of loss for the short-run period).

2.   Expansion in production capacity.

3. Increase in market share or in value of assets.

4. Diversification of activities.

5. Vertical and horizontal integration.

6. Continuous survival of the firm.

No single goal can express all of the complexities of the decision process. However, maximization of total profits (or loss minimization for the short-run period) is probably the principal goal of the management in many firms.

Listing all available projects according to priority ensures that a specific investment is justified and conforms with the firm's objectives.

1.4 CHARACTERISTICS OF OIL AND GAS RESOURCES

In addition to factors normally considered in capital investments, some of the characteristics of oil, gas, and many other natural resource development projects that may affect the results of economic analyses are as follows:

1. The long lead time from geologic discovery to the full use of the resource. At best, this may be several years; more often it is 8 to 12 years or longer.

2. The political and social environment in the region.

3. The nonrenewable nature of oil and gas resources.

4. Tax burden and special allowances customary in oil and gas accounting.

5. The heterogeneous nature of deposits (no two deposits are identical).

6. The finite quantity of commercial deposits.

7. Vulnerability to changing political and social conditions, especially in overseas countries (nationalization, takeovers, or shutdown of operations).

In the economic analyses of oil and gas projects these factors show up in terms of very long preproduction period (poorer returns on investment), a defined lifetime for the project, specific tax debits (royalties, property tax, severance tax) or credits (depletion allowance), and risk and uncertainty. How to incorporate these factors in economic evaluations is shown throughout this book.

1.5 DECISION MAKING ENVIRONMENT

The environment within which decision making takes place can be logically divided into three parts or states: certainty, risk, and uncertainty. Certainty exists when one can specify exactly what will happen during the period for which the decision is being made. Risk refers to a situation where one can specify a probability distribution over possible outcomes. Uncertainty refers to the condition when one cannot specify the relative likelihood of the outcomes.

People generally are more comfortable when they are dealing with certainty. The world seems more secure, and we plan ahead with confidence. But most of the decisions made in the oil and gas industry involve elements of risk and uncertainty. The drilling of an exploratory well is only one example where the uncertainties regarding both cost and the degree of success are so great that they cannot be ignored.

Oil and gas exploration and production are inherently probabilistic. By their very nature they include large elements of risk and uncertainty. The circumstances that lead to the generation of oil and gas are understood only in a general sense, but certainly they reflect the vagaries of the depositional environment. Migration of hydrocarbons into traps, and the creation of the traps themselves, are governed by processes that cannot be treated in a deterministic way except at an extremely simplistic level. The existence, or more particularly the location of traps, whether structural or stratigraphic, cannot be predicted with certainty. Even when a trap is successfully drilled, it may prove to be barren for no immediately discernible reason. Furthermore, the economic factors that ultimately affect the exploitation of resources are subject to capricious shifts that defy logical prediction. Thus the entire environment of oil and gas exploration and production is so permeated with chance factors that the business has often been called "the greatest gamble on earth."

The following are some questions (derived from Davis*) that must be answered sequentially before risk capital is exposed to petroleum exploitation are:

1. Will geophysical–geological studies locate a prospective anomaly on open acreage?

2. If so, will a wildcat well drilled into the prospect find hydrocarbons that can be produced at commercial rates?

3. Given a successful wildcat, will the delineation of the find prove the reserves necessary to justify expenditures for production facilities and the development of the discovery?

4. Finally, if the delineation phase is successful, will full exploitation of the discovery generate enough financial return consonant with profitability objectives?

Action on the prospect usually follows affirmative answers to the above questions.

There are no decision methods that eliminate or even reduce risk and uncertainty. The utilization of the methods discussed in this book will not reduce risk and uncertainty but rather will provide the necessary tools to evaluate, quantify, and understand risk and uncertainty so that the engineer, explorationist, and manager can devise a decision strategy to minimize the firm's exposure to risk and uncertainty.

---

*See the reference section at the end of each chapter for complete bibliographical data for text and illustrations quoted or adapted.

One of the pioneers of petroleum engineering, the late Everett L. DeGolyer put it this way:

> It takes luck to discover oil. Prospecting is like gin rummy. Luck enough will win but not skill alone. Best of all are luck and skill in proper proportions, but don't ask what the proportion should be. In the case of doubt, weight mine with luck.

The uncertainties in geology will always be as the Creator made them. How the problems of decision making in petroleum engineering are solved depend on how well engineers, geologists, geophysicists, and managers apply new ideas, knowledge, and technology. Petroleum exploitation is always an exciting and challenging game—a game of chance but also of change.

REFERENCES

Davis, L. F.: "Economic Judgement and Planning in North American Petroleum Exploration," *Journal of Petroleum Technology*, May 1968, pp. 467–474.

DeGolyer: "How Men Find Oil," *Fortune 97*, August 1948.

Rudawsky, O.: "Economic Feasibility Studies in Mineral and Energy Industries," Colorado School of Mines Mineral Industries Bulletin, Vol. 20, No. 3, May 1977.

2

ESTIMATION OF OIL AND GAS RESERVES

2.1 INTRODUCTION

Estimating oil and gas reserves is one of the most important phases in the economic evaluation of oil and gas drilling and producing ventures. The solutions to the problems of the petroleum engineer or explorationist depend on a comparison of estimated cost with anticipated result in terms of barrels of oil or cubic feet of gas. Careful and detailed prediction of expected performance permits the proper design and selection of equipment to maximize the economic benefits from the prospect.

Reliable reserve figures are most urgently needed early in the life of a reservoir when decisions are taken on large amounts of investment dollars. Unfortunately, only a minimal amount of information is available at this time. The amount and reliability of information increases as the producing property approaches its economic limit (Figure 2.1).

During the first period no wells have been drilled on the property. Reserve estimates are based on experience from similar pools or wells in the area or the same formation in other fields, and are usually expressed in barrels or cubic feet per acre. The estimates range all the way from the most pessimistic view, *AB* (property nonproductive) to the most optimistic view, *CD* (property very productive).

After one or more wells have been drilled and have been found to be productive, it becomes possible to estimate the ultimate recovery within narrower limits *EF* and *GH*. Estimates during this second period are on a volumetric basis and are usually expressed in barrels or cubic feet per acre-foot. Information available during this period include well logs, core-analysis data, bottom-hole samples, and subsurface maps.

The third period follows after sufficient production data become available. Material balance calculations may now be made. Volumetric estimates can be

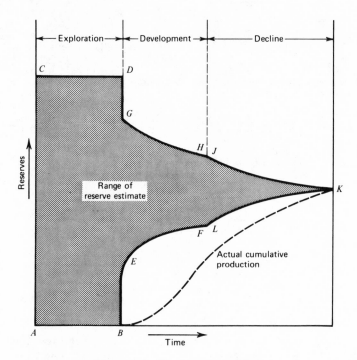

Figure 2.1 Reserves uncertainty

checked against material balance estimates and decline-curve trends. The type of production mechanism can also be determined. The lines *HJK* and *FLK* converge gradually to point *K*, which represents the true ultimate recovery. The broken curve *BK* represents the cumulative production.

Thus, primary recoverable oil and gas reserves can be determined by the following methods:

1. *Volumetric estimates* provided that information is available as to reservoir extent and thickness, reservoir porosity and water saturation, and oil or gas formation volume factor.

2. *Material balance estimates* provided that the production history and properties of produced fluids are known.

3. *Production decline curves* when extensive production history is available.

In addition, empirical methods have been developed for special reservoir cases.

The rest of this chapter deals mainly with volumetric estimates, material balances, and empirical correlation for oil and gas reservoirs. Production decline curves are discussed in Chapter 3.

2.2 CLASSIFICATION OF RESERVES

There is no agreement on the classification of oil and gas reserves. Standardized definitions for oil and gas supply indicators are not always used, estimation procedures differ, and professional judgment must be exercised in making resource estimates. Estimates of undiscovered oil and gas reserves that may eventually be found also differ greatly. The principles set forth for the U.S. Bureau of Mines and U.S. Geological Survey have been incorporated in a classification system intended for oil and gas (U.S. Federal Power Commission). A modification of these two classifications is shown diagrammatically in Fig. 2.2 and is explained by the accompanying definitions given by Meyer.

1. Resource—a concentration of naturally occurring solid or liquid petroleum or petroleum-like material, or natural gas, in or on the earth's crust in such a form [that] the economic extraction is currently or potentially feasible.

2. Discovered resources—resources whose location, quality, and quantity are known from drilling and geologic evidence, supported by engineering measurements.

3. Undiscovered resources—resources surmised to exist on the basis of broad geologic knowledge and theory.

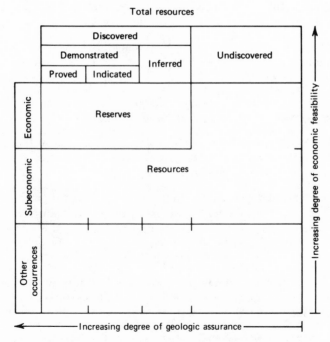

Figure 2.2 Resource classification diagram (after Meyer).

4. Reserve—that portion of the identified resource from which a usable mineral and energy commodity can be economically extracted at the time of estimation. Such commodities include, but are not necessarily restricted to, petroleum, condensate, natural gas, tar sands, and naturally occurring asphalt, without regard to mode of occurrence.

5. Proved reserves—material for which estimates of the quality and quantity have been computed from analyses and measurements from closely spaced and geologically well-known sample sites.

6. Indicated or probable reserves—material likely to be added in future years to proved reserves in discovered fields due to improved completion methods, and increased recovery efficiency by secondary or enhanced methods.

7. Demonstrated reserves—a collective term for the sum of proved and indicated reserves.

8. Inferred or possible reserves—material likely to be added in future years to proved reserves in discovered fields, estimated partly from drilling and production data and partly from extrapolation of geologic and engineering evidence over a reasonable area.

9. Discovered subeconomic resources—known resources not economically producible on the date of estimation. Such resources would include those that are too small or remote, at depths too great, or in water depths too great to be economic, or for which known producing technologies are not presently economic.

10. Hypothetical resources—undiscovered material that may reasonably be expected to exist in a known producing basin under known geologic conditions.

11. Speculative resources—undiscovered material that reasonably may be expected to exist in presently nonproductive basins.

12. Undiscovered subeconomic resources—that material in hypothetical and speculative deposits which, if found, would not be economic to produce, based upon existing technology.

13. Other occurrences—material not expected to become producible within a foreseeable period of time. For working purposes this period may be defined as 25 years from the date of estimation.

2.2.1 Proved Reserves

The Society of Petroleum Engineers of AIME (American Institute of Mechanical Engineers) has adopted the following definitions of proved reserves for property evaluation:

> Proved Reserves—The quantities of crude oil, natural gas and natural gas liquids which geological and engineering data demonstrate with reasonable certainty to be recoverable in the future from known oil and gas reservoirs under existing economic

and operating conditions. They represent strictly technical judgments, and are not knowingly influenced by attitudes of conservatism or optimism.

Undrilled Acreage—Both drilled and undrilled acreage of proved reservoirs are considered in the estimates of the proved reserves. The proved reserves of the undrilled acreage are limited to those drilling units immediately adjacent to the developed areas, which are virtually certain of productive development, except where the geological information on the producing formations insures continuity across other undrilled acreage.

Fluid Injection—Additional reserves to be obtained through the application of fluid injection or other improved recovery techniques for supplementing the natural forces and mechanisms of primary recovery are included as "proved" only after testing by a pilot project or after the operation of an installed program has confirmed that increased recovery will be achieved.

When evaluating an individual property in an existing oil or gas field, the proved reserves within the framework of the above definition are those quantities indicated to be recoverable commercially from the subject property at current prices and costs, under existing regulatory practices, and with conventional methods and equipment. Depending on their development or producing status, these proved reserves are further subdivided into the following categories:

1. Proved Developed Reserves—Proved reserves to be recovered through existing wells and with existing facilities;

    a. Proved Developed Producing Reserves—Proved developed reserves to be produced from completion interval(s) open to production in existing wells;

    b. Proved Developed Nonproducing Reserves—Proved developed reserves behind the casing of existing wells or at minor depths below the present bottom of such wells which are expected to be produced through these wells in the predictable future. The development cost of such reserves should be relatively small compared to the cost of a new well.

2. Proved Undeveloped Reserves—Proved reserves to be recovered from new wells on undrilled acreage or from existing wells requiring a relatively major expenditure for recompletion or new facilities for fluid injection.

2.2.2 Probable Reserves

In an evaluation of an oil or gas property, estimates of reserves that are not definitely proven are usually needed in making economic decisions.

Probable Primary Reserves are those that have not been proved by production at a commercial rate of flow and must be proved by additional drilling and testing.

Probable Secondary Reserves are thought to exist in a reservoir by virtue of past production performance or core, well log, or reservoir data, but the reservoir itself has not been subjected to successful secondary recovery operations.

2.2.3 Possible Reserves

These are reserves that may be present in nonproducing regions where geological or other reservoir information suggests that producible hydrocarbons may be present.

Possible Primary Reserves are reserves that may exist but for which available information does not support a higher classification.

Possible Secondary Reserves are secondary reserves from reservoirs that appear to be suited for secondary recovery operations but for which available information will not support a higher classification.

2.3 CLASSIFICATION OF OIL AND GAS

Hydrocarbons compounds, when produced through wells, may be in the liquid phase (liquid hydrocarbons) or in the gaseous phase (natural hydrocarbon gases).

2.3.1 Liquid Hydrocarbons

These are subdivided into (Arps 1962):

1. Crude oil—consisting primarily of intermediate and heavy hydrocarbon compounds and occurring in a natural state under reservoir conditions. They are produced or recovered as a liquid by ordinary field separating equipment.

2. Natural-gas liquids—liquid hydrocarbons consisting primarily of light and intermediate hydrocarbons and occurring as a free-gas phase or in solution with the crude oil in an oil reservoir. They are recoverable as liquids by condensation or absorption in field separators, scrubbers, gasoline plants, or recycling plants. Natural-gas liquids may be subdivided further:

   (a) Condensate—low-vapor-pressure products recoverable by ordinary field separation equipment. This is often combined and recorded with crude oil.
   (b) Natural gasoline—intermediate-vapor-pressure products recoverable by special separation equipment or gasoline plants.
   (c) Liquefied petroleum gases—primarily high-vapor-pressure products such as butane, propane, and ethane, recoverable in specially equipped gasoline plants and which can be maintained in the liquid phase only under substantial pressure.

2.3.2 Natural Hydrocarbon Gases

These primarily consist of the lighter paraffin hydrocarbons and are subdivided into (Arps 1962):

1. Nonassociated gas—gaseous hydrocarbons occurring as a free-gas phase under original conditions in the reservoir, which is not commercially productive of crude oil.

2. Associated gas—gaseous hydrocarbons occurring as a free gas phase under original reservoir conditions, in contact with a commercially productive crude-oil reservoir.

3. Dissolved or solution gas—gaseous hydrocarbons occurring under original

reservoir conditions in solution with crude oil in a commercially productive crude-oil reservoir.

4. Injected gas—gaseous hydrocarbons that have been injected in underground reservoirs for pressure maintenance or storage purposes.

2.4 VOLUMETRIC ESTIMATES

Long before a well is drilled, the geologist or petroleum engineer makes volumetric estimates of the oil or gas that might exist in the reservoir, if the reservoir exists. These early estimates are usually obtained by analogy with similar and/or offset properties having similar geological and other reservoir conditions. These estimates, usually expressed in terms of recoverable barrels or thousand standard cubic feet per acre, may be the justification for spending a lot of money to acquire the property and drill the first well.

To obtain the total recoverable oil or gas from the property, estimates of possible productive acres will have to be made. These areal estimates may be obtained from geological maps using surrounding wells for control or from geophysical maps. After the first well is drilled, much more information on the reservoir is available. Surfaces that can be sampled by a well include (a) structure tops, (b) isopachs, (c) porosities, (d) permeabilities, (e) saturations, (f) pressures, (g) contacts, and (h) discontinuities.

Usually the way the samples on a common surface are displayed is through the art of contouring. Figure 2.3 shows the contouring of some sampled surfaces.

2.4.1 Computation of Reservoir Volume

When sufficient subsurface control is available, the oil- or gas-bearing net pay volume of a reservoir may be computed by planimetering or numerical integration.

Planimetering

A planimeter is a device that integrates surface areas. By planimetering the area enclosed by each contour on an isopach map, it is possible to compute total

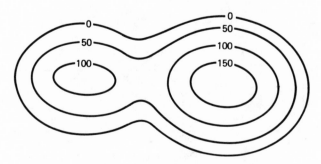

Figure 2.3 Isopach map; contour interval: 50 feet.

volume. Reservoir volume may be computed by (a) using an acre-feet diagram, (b) slicing the reservoir horizontally and summing volumes, or (c) slicing the reservoir in the vertical direction and summing volumes. These methods will be illustrated using an isopach map (Fig. 2.3).

Acre-Feet Diagram

From the subsurface data a geological map (Fig. 2.3) is prepared. The total area enclosed by each contour is then planimetered and plotted as abscissa on an acre-feet diagram (Fig. 2.4) against the corresponding isopach as the ordinate. After connecting the observed points the volume, which is represented by the area under the curve, is determined by planimetering from the acre-feet diagram, by using formulas like Simpson's rule, trapezoidal rule, or pyramidal rule, or by numerical integration.

This technique may be subject to errors due to how the curve is drawn through the points and reservoir-wide averaging of data. It is not well suited to determining reservoir volumes for many individual tracts such as exist in unitization efforts.

Horizontal Slices

The geological map can be conceptually sliced horizontally (Fig. 2.5). The areas enclosed within contours become bases and tops for trapezoidal volumes, frustums of pyramids, frustums of cones, or spherical segments. The degree of accuracy in using this method depends on how well the models match the actual shape of the reservoir. Reservoir-wide averaging of data will also give rise to error in volume computation. Volumes of individual tracts are not easily determined.

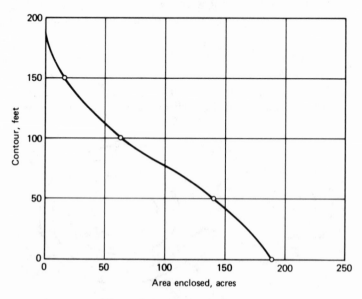

Figure 2.4 Acre-feet diagram.

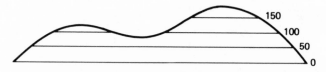

Figure 2.5   Horizontal slicing of isopach map.

Vertical Slices

The geological map can be conceptually sliced vertically (Fig. 2.6). The contour lines now become the sides of vertical cylinders or various annular-shaped volumes. The error from reservoir-wide averaging of data is reduced. The vertical planes may be made to coincide with the division of ownership in the reservoir. However, the models and formulas used may not match the reservoir accurately.

Some Useful Mensuration Formulas

Trapezoidal Rule

$$V = \tfrac{1}{2} h \, (A_0 + 2A_1 + 2A_2 + 2A_3 + \cdots + 2A_{n-1} + A_n) \qquad (2.1)$$

Simpson's Rule

$$V = \tfrac{1}{3} h \, (A_0 + 4A_1 + 2A_2 + 4A_3 + 2A_4 + \cdots + 2A_{n-2}$$
$$+ 4A_{n-1} + A_n) \qquad (2.2)$$

This formula can be used if the number of contour intervals, n is even (i.e., an odd number of contours). It is mathematically correct only for plane surfaces.

Trapezoidal Volume

$$V = \tfrac{1}{2} h \, (A_1 + A_2) \qquad (2.3)$$

A_1 = area of lower base, A_2 = area of upper base, and h = altitude.

Pryamid or Cone

$$V = \tfrac{1}{3} (\text{area of base}) \times (\text{altitude}) \qquad (2.4)$$

Frustum of Pyramid or Frustum of Cone

$$V = \tfrac{1}{3} h \, (A_1 + A_2 + \sqrt{A_1 \, A_2}) \qquad (2.5)$$

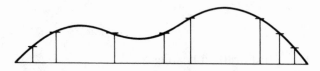

Figure 2.6   Vertical slicing of isopach map.

Spherical Segment on One Base

$$V = \tfrac{1}{6} \pi h \, (3r_1^2 + h^2)$$ (2.6)

Spherical Segment on Two Bases

$$V = \tfrac{1}{6} \pi h \, (3r_1^2 + 3r_2^2 + h^2)$$ (2.7)

Circular Cylinder

$$V = A_1 h = \pi R^2 h$$ (2.8)

Example 2.1.   The isopach map shown in Fig. 2.3 may be assumed to represent an oil field. By planimetering, the following areas are obtained for the contour lines:

h (ft)	Area (acres)
0	190
50	140
100	60
150	20

Determine the bulk volume of the reservoir.

Solution
Trapezoidal rule gives

$$V = \tfrac{50}{2} \, (190 + 2 \times 140 + 2 \times 60 + 20) = 15{,}250 \text{ acre-ft}$$

For frustum of a cone:

$$V_{0-50} = \tfrac{50}{3} \, (190 + 140 + \sqrt{190 \times 140}) = \; 8{,}218$$
$$V_{50-100} = \tfrac{50}{3} \, (140 + 60 + \sqrt{140 \times 60}) = \; 4{,}861$$
$$V_{100-150} = \tfrac{50}{3} \, (60 + 20 + \sqrt{60 \times 20}) = \; 1{,}911$$
$$\text{Total} \; = 14{,}990 \text{ acre-ft}$$

All methods based on planimetered areas require using average values of reservoir data over the planimetered areas. The hydrocarbons-in-place are usually calculated by:

$$\text{Hydrocarbons-in-place} = V_B \cdot \phi \cdot S_{hc}$$ (2.9)

where

V_B = volume of hydrocarbon bearing rock

ϕ = porosity (fraction of the rock that is void space)

S_{hc} = oil saturation (fraction of the void space that contains hydrocarbons)

Since several surfaces are involved, for example isopach, isoporosity, and iso-saturation, all of these surfaces are averaged individually over the planimetered area. As shown by Eq. 2.9, the product of integrals is used. This may introduce considerable amount of error in reserve calculations.

Equation 2.9 can be written for recoverable oil in place as:

$$N = 7,758Ah \, \phi \frac{(1 - S_{wi})}{B_{oi}} E_R \qquad (2.10)$$

and for recoverable gas in place as:

$$G = 43,560Ah \, \phi \frac{(1 - S_w)}{B_{gi}} E_R \qquad (2.11)$$

where

N = oil-in-place, stock tank barrels (STB)

G = gas-in-place, standard cubic feet (scf)

7758 = bbl/acre-ft

43,560 = ft^2/acre-ft

A = area, acres

h = net formation thickness, feet

ϕ = porosity, fraction

S_{wi} = initial water saturation, fraction

B_{oi} = initial oil formation volume factor, reservoir barrels per stock tank barrel (res bbl/STB)

B_{gi} = initial gas formation volume factor, reservoir cubic feet per standard cubic foot (res ft^3/scf)

E_R = recovery factor (fraction of the oil or gas in place that will probably be recovered)

Errors in these reservoir parameters will of course yield corresponding errors in the value of computed oil- or gas-in-place. Typical levels of errors in volumetric calculations are as follows:

A	10% possible error
h	10% possible error
ϕ	15% possible error

$$S_{wi} \qquad 5\% \text{ possible error}$$
$$E_R \qquad 10\% \text{ possible error}$$

Numerical Integration

More accurate mathematical statements of the volumetric estimates are

$$N = 7758 \, E_R \sum_{j=1}^{n} \frac{(A_j)(h_j)(\phi_j)(1 - S_{wj})}{B_{oij}} \tag{2.12}$$

and

$$G = 43{,}560 \, E_R \sum_{j=1}^{n} \frac{(A_j)(h_j)(\phi_j)(1 - S_{wj})}{B_{gij}} \tag{2.13}$$

Equations 2.12 and 2.13 suggest that discrete values for each parameter must be picked from each contoured surface. Then the product of the discrete values is summed together. *The integral of products is not equal to the product of integrals.* The error introduced by reservoir-wide averaging is greatly reduced. The only requirement is a contour map for each sampled surface.

Numerical integration may be performed by hand or on a computer. The job is faster, less tedious, and more accurate when performed by a computer. Also much smaller sample areas may be justified when using a computer.

A contour map is required for each independent variable. Some of the maps needed are:

1. Structure map on top of reservoir.

2. Isopach of gross reservoir thickness or structure map on the base of the reservoir.

3. Structure map on unconformity.

4. Isoporosity map.

5. Net-to-gross ratio map.

6. Water−oil contact map.

7. Gas−oil contact map.

Examples of some of these maps are given in Figs. 2.7 to 2.9.

When all required maps have been obtained, an arbitrary sample size is selected. This is based on the desired accuracy and the variability of the contoured surfaces. A transparent grid overlay is prepared as shown in Fig. 2.10. Gross thickness is picked from the gross pay isopach. Gross oil and/or gas thicknesses can be calculated at each sample point by considering water−oil contact, gas−oil contact, and unconformity. The gross volumes are then modified by the net-to-gross ratio, porosity, and water saturation. Sample volumes can be summed to lease, tract, or field volumes.

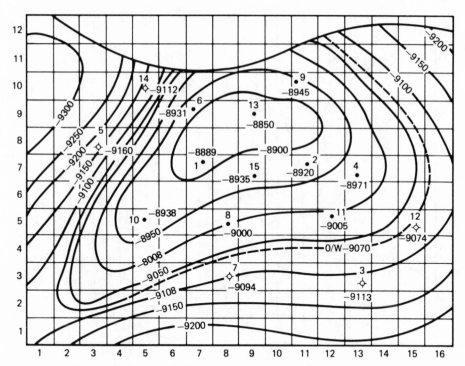

Figure 2.7 Structure top of zone 1; contour interval: 50 feet. Legend: ● oil well; ✧ dry well (courtesy of Core Laboratories, Inc.)

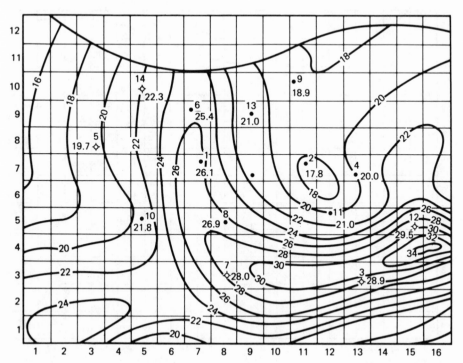

Figure 2.8 Isoporosity of zone 1, contour interval: 2%. Legend: ● oil well; ✧ dry well (courtesy of Core Laboratories, Inc.)

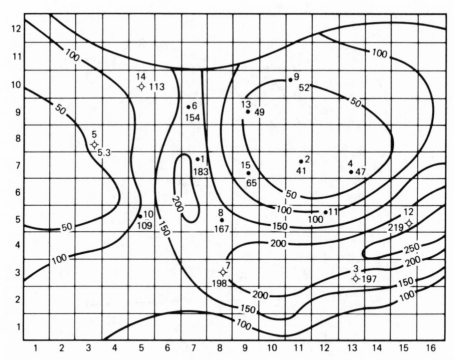

Figure 2.9 Horizontal permeability of zone 1, contour interval: 50 md. Legend: ● oil well; ✧ dry hole (courtesy of Core Laboratories, Inc.)

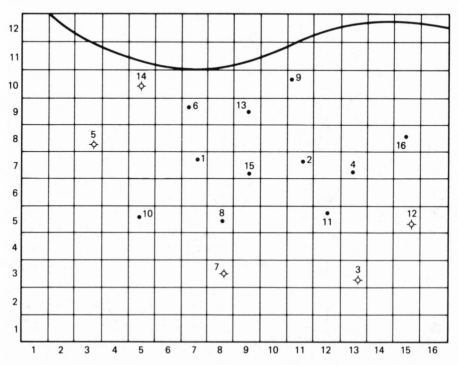

Figure 2.10 Grid system of zone 1 (courtesy of Core Laboratories, Inc.)

2.4.2 Volumetric Estimates of Oil Reserves

The oil recovery from a given field is a function of the prevailing drive mechanism, reservoir configuration, fluids involved, reservoir type, and prevailing reservoir pressure and temperature. Equation 2.10 or any of its modified forms may be used to determine the recoverable oil reserves. The acre-feet of oil reservoir rock, Ah, are determined from an isopach map. Water saturation information is determined from well logs, core saturation studies in the laboratory, or from information on a similar reservoir. Porosity is found chiefly by the analysis of well logs and cores. The form that the recovery factor takes will depend upon the producing mechanism of the reservoir.

Example 2.2. A well has been drilled into an oil reservoir. Well logs indicate that the reservoir has a thickness of 105 feet, average porosity of 18.5 percent, and average water saturation of 25 percent. Analysis of offset wells production in the area indicate that this well will drain 160 acres with a recovery efficiency of 15 percent, and the initial oil formation volume factor is 1.25. What are the recoverable oil reserves?

Solution

$$N = \frac{(7,758)(160)(105)(0.185)(1 - 0.25)(0.15)}{1.25}$$

$$= 2,170,068 \text{ stock tank barrels (STB)}$$

Solution Gas Drive (Depletion-Type) Reservoirs

The most common assumption is that an oil property is producing by the solution gas drive mechanism. For this reservoir type, Eq. 2.10 can be written as:

$$N = 7,758Ah \, \phi \left(\frac{1 - S_w}{B_{oi}} - \frac{1 - S_w - S_{gr}}{B_{oa}} \right) \qquad (2.14)$$

where

B_{oi} = oil formation volume factor at initial conditions (at or below the bubble point pressure)

B_{oa} = oil formation volume factor at abandonment pressure, reservoir bbl/STB

S_{gr} = residual gas saturation in the reservoir at abandonment pressure (fraction of pore space)

All other parameters are as defined before. Equation 2.14 has the recovery efficiency included in the terms in brackets. The primary difficulty in the use of Eq. 2.14 is in the determination of a value for S_{gr}. Smith has suggested a value for S_{gr} of 0.25, if the initial solution gas-oil ratio is 400 to 500 scf/STB, and the oil gravity is between 30 and 40°API (American Petroleum Institute). The S_{gr}

increases (decreases) approximately 0.01 for every 3°API increase (decrease) in oil gravity. A 50 percent decrease in solution gas–oil ratio would decrease the S_{gr} by 0.05. A doubling of the solution gas–oil ratio, or complete lack of shaliness in a loosely cemented sandstone could increase the S_{gr} by as much as 0.10.

Example 2.3. Consider Example 2.2 again. Based on gas–oil ratio and oil gravity from offset well production, the residual gas saturation has been estimated as 30 percent and the oil formation volume factor as 1.15 at abandonment. What are the recoverable oil reserves at abandonments?

Solution

$$N = (7,758)(160)(105)(0.185) \left(\frac{1 - 0.25}{1.25} - \frac{1 - 0.25 - 0.30}{1.15} \right)$$

$$= 5,032,041 \text{ STB}$$

Arps has presented a study of recovery factors for solution gas drive reservoirs. Table 2.1 summarizes the results of this study. In cases where no detailed data are available on the characteristics of the reservoir rock and fluids, this table has been found to be very helpful in estimating the possible range of solution gas drive recovery factors.

Wahl, Mullins, and Elfrink have prepared nine charts (Figs. 2.11 to 2.19) for estimating oil recovery from solution gas drive reservoirs. These charts give ultimate oil recovery, as a percent of the residual oil originally in place. The

TABLE 2.1
Primary Recovery Factors for Depletion-Type Reservoirs

Soln GOR	Oil Gravity °API	Sand or Sandstones			Limestone, Dolomite, or Chert		
		Maximum	Average	Minimum	Maximum	Average	Minimum
60	15	12.8	8.6	2.6	28.0	4.4	0.6
	30	21.3	15.2	8.7	32.8	9.9	2.9
	50	34.2	24.8	16.9	39.0	18.6	8.0
200	15	13.3	8.8	3.3	27.5	4.5	0.9
	30	22.2	15.2	8.4	32.3	9.8	2.6
	50	37.4	26.4	17.6	39.8	19.3	7.4
600	15	18.0	11.3	6.0	26.6	6.9	1.9
	30	24.3	15.1	8.4	30.0	9.6	(2.5)
	50	35.6	23.0	13.8	36.1	15.1	4.3
1000	15	—	—	—	—	—	—
	30	34.4	21.2	12.6	32.6	13.2	(4.0)
	50	33.7	20.2	11.6	31.8	12.0	(3.1)
2000	15	—	—	—	—	—	—
	30	—	—	—	—	—	—
	50	40.7	24.8	15.6	32.8	(14.5)	(5.0)

Source: Arps 1962.

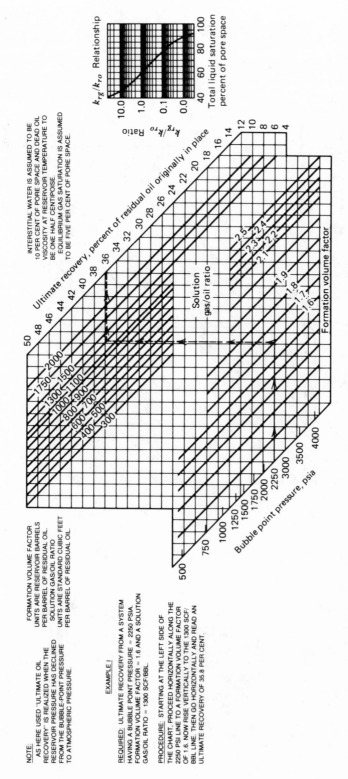

Figure 2.11 Chart for estimating ultimate recovery from solution gas-drive reservoirs, $S_{wc} = 10\%$, $\mu_0 = 0.5$ cp. (courtesy of SPE of AIME).

23

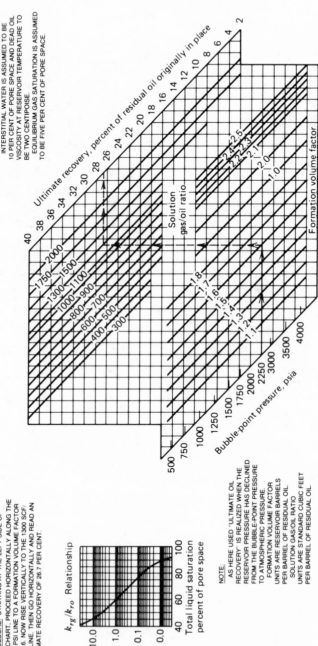

Figure 2.12 Chart for estimating ultimate recovery from solution gas-drive reservoirs, $S_{wc} = 10\%$, $\mu_o = 2$ cp. (courtesy of SPE of AIME).

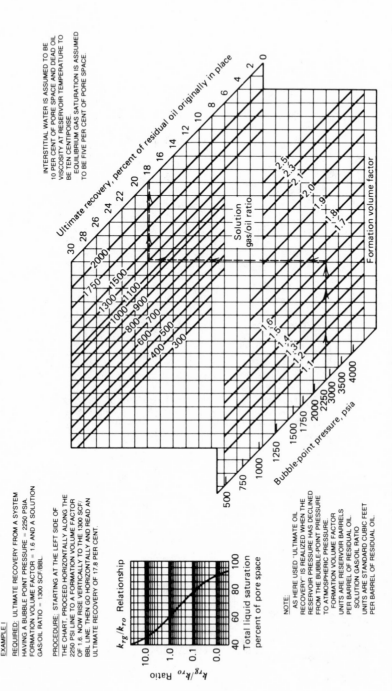

Ultimate recovery, percent of residual oil originally in place

Solution gas/oil ratio

Formation volume factor

Bubble-point pressure, psia

k_{rg}/k_{ro} Relationship

k_{rg}/k_{ro} Ratio

Total liquid saturation percent of pore space

EXAMPLE I

REQUIRED: ULTIMATE RECOVERY FROM A SYSTEM
HAVING A BUBBLE POINT PRESSURE = 2250 PSIA.
FORMATION VOLUME FACTOR = 1.6 AND A SOLUTION
GAS/OIL RATIO = 1300 SCF/BBL.

PROCEDURE: STARTING AT THE LEFT SIDE OF
THE CHART, PROCEED HORIZONTALLY ALONG THE
2250 PSI LINE TO A FORMATION VOLUME FACTOR
OF 1.6. NOW RISE VERTICALLY TO THE 1300 SCF/
BBL LINE. THEN GO HORIZONTALLY AND READ AN
ULTIMATE RECOVERY OF 17.8 PER CENT.

NOTE:

AS HERE USED "ULTIMATE OIL
RECOVERY" IS REALIZED WHEN THE
RESERVOIR PRESSURE HAS DECLINED
FROM THE BUBBLE-POINT PRESSURE
TO ATMOSPHERIC PRESSURE.
FORMATION VOLUME FACTOR
UNITS ARE RESERVOIR BARRELS
PER BARREL OF RESIDUAL OIL.
SOLUTION GAS/OIL RATIO
UNITS ARE STANDARD CUBIC FEET
PER BARREL OF RESIDUAL OIL.

Figure 2.13 Chart for estimating ultimate recovery from solution gas-drive reservoirs, $S_{wc} = 10\%$, $\mu_0 = 10$ cp. (courtesy of SPE of AIME).

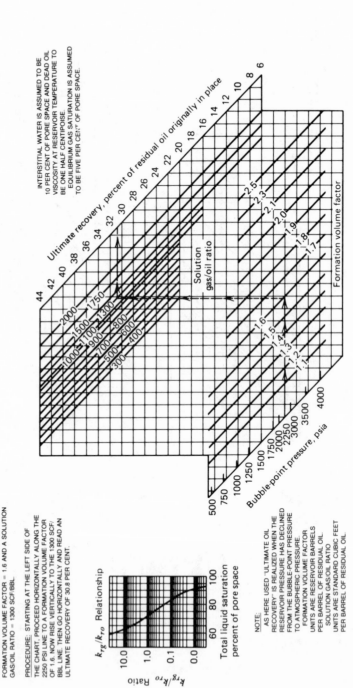

Figure 2.14 Chart for estimating ultimate recovery from solution gas-drive reservoirs, $S_{wc} = 30\%$, $\mu_0 = 0.5$ cp. (courtesy of SPE of AIME).

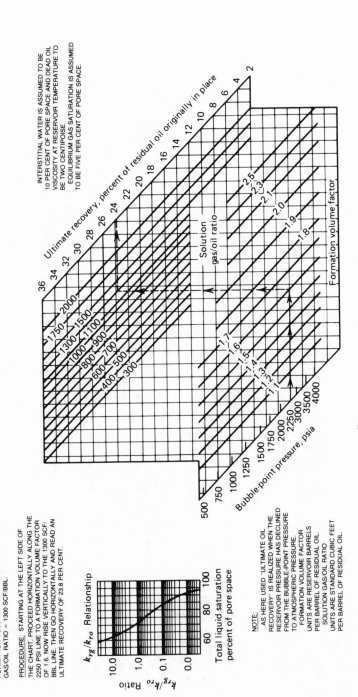

Figure 2.15 Chart for estimating ultimate recovery from solution gas-drive reservoirs, $S_{wc} = 30\%$, $\mu_0 = 2$ cp. (courtesy of SPE of AIME).

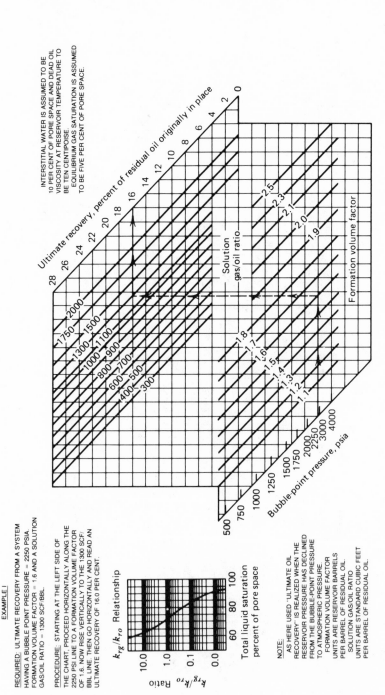

Figure 2.16 Chart for estimating ultimate recovery from solution gas-drive reservoirs, $S_{wc} = 30\%$, $\mu_0 = 10$ cp. (courtesy of SPE of AIME).

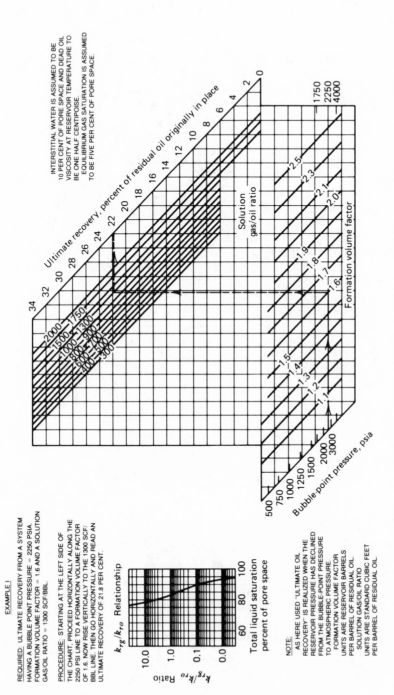

EXAMPLE I

REQUIRED: ULTIMATE RECOVERY FROM A SYSTEM HAVING A BUBBLE POINT PRESSURE = 2250 PSIA, FORMATION VOLUME FACTOR = 1.6 AND A SOLUTION GAS/OIL RATIO = 1300 SCF/BBL.

PROCEDURE: STARTING AT THE LEFT SIDE OF THE CHART, PROCEED HORIZONTALLY ALONG THE 2250 PSI LINE TO A FORMATION VOLUME FACTOR OF 1.6. NOW RISE VERTICALLY TO THE 1300 SCF/ BBL LINE. THEN GO HORIZONTALLY AND READ AN ULTIMATE RECOVERY OF 21.8 PER CENT.

k_{rg}/k_{ro} Relationship

NOTE:
AS HERE USED "ULTIMATE OIL RECOVERY" IS REALIZED WHEN THE RESERVOIR PRESSURE HAS DECLINED FROM THE BUBBLE-POINT PRESSURE TO ATMOSPHERIC PRESSURE. FORMATION VOLUME FACTOR UNITS ARE RESERVOIR BARRELS PER BARREL OF RESIDUAL OIL. SOLUTION GAS/OIL RATIO UNITS ARE STANDARD CUBIC FEET PER BARREL OF RESIDUAL OIL.

INTERSTITIAL WATER IS ASSUMED TO BE 10 PER CENT OF PORE SPACE AND DEAD OIL VISCOSITY AT RESERVOIR TEMPERATURE TO BE ONE HALF CENTIPOISE. EQUILIBRIUM GAS SATURATION IS ASSUMED TO BE FIVE PER CENT OF PORE SPACE.

Figure 2.17 Chart for estimating ultimate recovery from solution gas-drive reservoirs, $S_{wc} = 50\%$, $\mu_0 = 0.5$ cp. (courtesy of SPE of AIME).

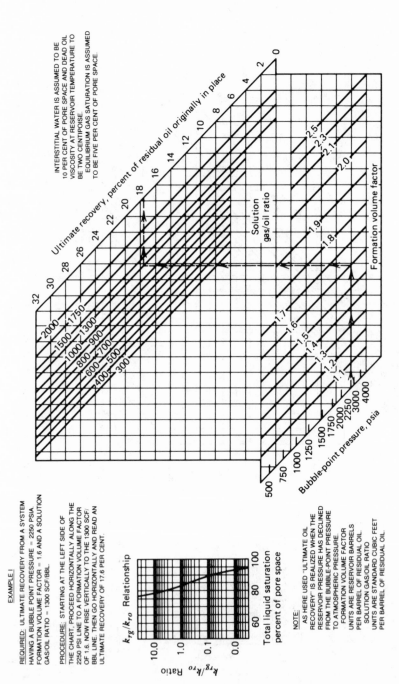

Figure 2.18 Chart for estimating ultimate recovery from solution gas-drive reservoirs, $S_{wc} = 50\%$, $\mu_0 = 2$ cp. (courtesy of SPE of AIME).

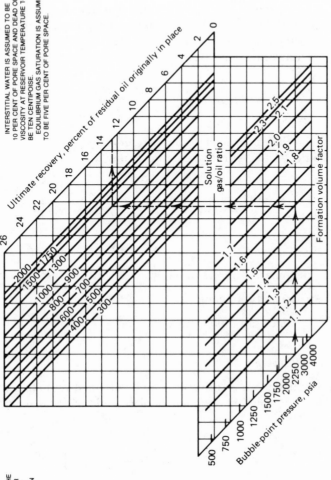

EXAMPLE I

REQUIRED: ULTIMATE RECOVERY FROM A SYSTEM
HAVING A BUBBLE POINT PRESSURE = 2250 PSIA.
FORMATION VOLUME FACTOR = 1.6 AND A SOLUTION
GAS/OIL RATIO = 1300 SCF/BBL.

PROCEDURE: STARTING AT THE LEFT SIDE OF
THE CHART, PROCEED HORIZONTALLY ALONG THE
2250 PSI LINE TO A FORMATION VOLUME FACTOR
OF 1.6. NOW RISE VERTICALLY TO THE 1300 SCF/
BBL LINE. THEN GO HORIZONTALLY AND READ AN
ULTIMATE RECOVERY OF 12.7 PER CENT.

k_{rg}/k_{ro} Relationship

NOTE:

AS HERE USED "ULTIMATE OIL
RECOVERY" IS REALIZED WHEN THE
RESERVOIR PRESSURE HAS DECLINED
FROM THE BUBBLE-POINT PRESSURE
TO ATMOSPHERIC PRESSURE
  FORMATION VOLUME FACTOR
  UNITS ARE RESERVOIR BARRELS
  PER BARREL OF RESIDUAL OIL.
  SOLUTION GAS/OIL RATIO
  UNITS ARE STANDARD CUBIC FEET
  PER BARREL OF RESIDUAL OIL.

Figure 2.19 Chart for estimating ultimate recovery from solution gas-drive reservoirs, $S_{wc} = 50\%$, $\mu_0 = 10$ cp. (courtesy of SPE of AIME).

ultimate oil recovery has been restricted to that oil recovered as the reservoir pressure declined from the bubble-point to atmospheric pressure. If the reservoir fluid is undersaturated at the initial reservoir pressure, the total recovery would be the sum of (1) the recovery due to fluid expansion as the reservoir pressure decreased from the initial value to the bubble-point pressure, and (2) the ultimate oil recovery estimated by means of one of the charts. Above the bubble-point pressure, the oil recovery is given by

$$N_P = \frac{7758Ah\phi\,(p_i - p_b)\,C_{oe}}{B_{oi}\,[1 + C_{oe}\,(p_i - p_b)]} \qquad (2.15)$$

N_P is the oil produced between the initial reservoir pressure and the bubble point pressure and C_{oe} is the effective compressibility defined by

$$C_{oe} = C_o + \left(\frac{S_w}{1 - S_w}\right) C_w + \left[\frac{1 - \phi}{\phi(1 - S_w)}\right] C_f \qquad (2.16)$$

where

C_o, C_w, C_f = compressibilities of oil, water, and rock, respectively, vol/vol-psi

ϕ = porosity, fraction

S_w = water saturation, fraction

Water Drive Reservoirs

The recovery under water drive may be determined by taking the difference between the stock tank oil originally in place and the residual stock tank oil at abandonment time.

$$N = 7758Ah\phi \left(\frac{1 - S_w}{B_{oi}} - \frac{S_{or}}{B_{oa}}\right) \qquad (2.17)$$

where S_{or} is the residual oil saturation at abandonment time as a fraction of pore space.

The residual oil saturation at abandonment time may be found by actually waterflooding cores in the laboratory under simulated reservoir conditions (flood-pot tests). Another method of estimating S_{or} is to use the oil saturations determined by core analysis after multiplying by the oil formation volume factor at abandonment (B_{oa}) as the residual oil saturation in the reservoir after water flooding. This is based on the assumption that water from the drilling mud invades the pay section just ahead of the core bit in a manner similar to the water displacement process in the reservoir. Approximations for S_{or} are given by Arps in Table 2.2. The deviations from these average saturations are given in Table 2.3. According to Tables 2.2 and 2.3, the residual oil saturation for a water drive

TABLE 2.2
Residual Oil Saturation (Water Drive Reservoirs)

Reservoir Oil Viscosity (cp)	Residual Oil Saturation (Percent of Pore Space)
0.2	30
0.5	32
1.0	34.5
2.0	37
5.0	40.5
10.0	43.5
20.0	46.5

Source: American Petroleum Institute.

TABLE 2.3
Deviation of S_{or} From Viscosity Trend
(Water Drive Reservoirs)

Average Reservoir Permeability (md)	Deviation of S_{or} (%)
50	+12
100	+ 9
200	+ 6
500	+ 2
1,000	− 1
2,000	− 4.5
5,000	− 8.5

Source: American Petroleum Institute.

reservoir containing 2 cp oil and having an average permeability of 200 md can be estimated at 37 + 6, or 43 percent of pore space.

Gas-Cap Drive Reservoirs

Equation (2.17) may be applied to gas-cap drive reservoirs. As a first estimate, Smith has suggested using values of S_{or} in the range of 10 to 20 percent. To obtain reliable estimates of S_{or}, it is usually necessary to make frontal displacement calculations using the Buckley−Leverett method.

API Correlation

The API subcommittee on recovery efficiency has published results of statistical study of recovery efficiencies. For this study the subcommittee used case histories on some 312 producing oil reservoirs including water drive, gas cap drive, gravity drainage, and solution gas drive. The range of ultimate recoveries for each of the predominant mechanisms is shown in Table 2.4.

The subcommittee presented correlations for water drive reservoirs and solution gas drive reservoirs.

TABLE 2.4
Ultimate Recoveries of Oil

	Sand and Sandstone			Limestone, Dolomite, and Other		
	Minimum	Medium	Maximum	Minimum	Medium	Maximum
Water drive						
BAF, bbl/AF	155	571	1,641	6	172	1422
E_R, %	27.8	51.1	86.7	6.3	43.6	80.5
S_{gr}, fraction	0.114	0.327	0.635	0.247	0.421	0.908
Solution gas drive without supplemental drives						
BAF, bbl/AF	47	154	534	20	88	187
E_R, %	9.5	21.3	46.0	15.5	17.6	20.7
S_{gr}, fraction	0.130	0.229	0.382	0.169	0.267	0.447
Solution gas drive with supplemental drives						
BAF, bbl/AF	109	227	820	32	120	464
E_R, %	13.1	28.4	57.9	9.0	21.8	48.1
S_{gr}, fraction	0.077	0.255	0.435	0.112	0.260	0.426
Gas cap drive						
BAF, bbl/AF	68	289	864			
E_R, %	15.8	32.5	67.0	Combined with sand		
S_{gr}, fraction	0.223	0.271	0.571	and sandstone		
Gravity drainage						
BAF, bbl/AF	250	696	1,124			
E_R, %	16	57.2	63.8	Data not available		
S_{gr}, fraction	0.151	0.377	0.654			

Source: American Petroleum Institute.

For *water drive reservoirs* (sands and sandstones) the best equation for recovery factor is

$$BAF = (4259) \left\{ \frac{\phi(1 - S_w)}{B_{oi}} \right\}^{+1.0422} \left(\frac{k\mu_{wi}}{\mu_{oi}} \right)^{+0.0770}$$

$$\cdot (S_w)^{-0.1903} \left(\frac{p_i}{p_a} \right)^{-0.2159} \text{ bbl/acre-ft} \qquad (2.18)$$

or

$$E_R = (54.898) \left\{ \frac{\phi(1 - S_w)}{B_{oi}} \right\}^{+0.0422} \left(\frac{k\mu_{wi}}{\mu_{oi}} \right)^{+0.0770}$$

$$\cdot (S_w)^{-0.1903} \left(\frac{p_i}{p_a} \right)^{-0.2159} \% \qquad (2.19)$$

For *solution gas drive reservoirs* (sands, sandstones, and carbonate rocks) the best equation for recovery factor below the bubble-point pressure is

$$BAF = (3244) \left\{ \frac{\phi(1 - S_w)}{B_{ob}} \right\}^{+0.1611} \left(\frac{k}{\mu_{ob}} \right)^{+0.0979}$$

$$\cdot (S_w)^{+0.3722} \left(\frac{p_b}{p_a} \right)^{+0.1741} \quad \text{bbl/acre-ft} \quad (2.20)$$

or

$$E_R = (41.815) \left\{ \frac{\phi(1 - S_w)}{B_{ob}} \right\}^{+0.1611} \left(\frac{k}{\mu_{ob}} \right)^{+0.0979}$$

$$\cdot (S_w)^{+0.3722} \left(\frac{p_b}{p_a} \right)^{+0.1741} \quad \% \quad (2.21)$$

Figures 2.20 and 2.21 show nomographs for solving Eqs. 2.19 and 2.21. The nomenclature for these equations is as follows:

BAF = ultimate recovery in stock tank barrels of oil per acre foot of net pay

ϕ = effective porosity, fraction

S_w = interstitial water saturation, fraction

B_{oi} = oil formation volume factor at initial conditions, res vol/stock tank vol

B_{ob} = oil formation volume factor at bubble point

k = arithmetic average of absolute permeability, darcys

μ_{wi} = viscosity of reservoir water at initial conditions, centipoises

μ_{oi} = viscosity of reservoir oil at initial conditions, centipoises

μ_{ob} = viscosity of reservoir oil at bubble point, centipoises

p_i = pressure at initial conditions, psig

p_b = bubble-point pressure, psig

p_a = pressure at abandonment, psig

2.4.3 Volumetric Estimates of Gas Reserves

The standard cubic feet (scf) of gas initially in place (G) is simply the product of three factors: the reservoir pore volume, initial gas saturation, and a volume ratio (initial gas formation volume factor) that converts reservoir volumes to volumes at standard, or base, conditions (normally 60°F and 14.7 psia). These factors are

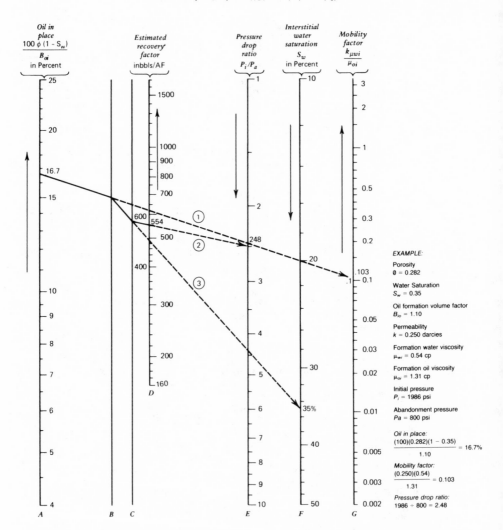

Based on: $4259 \times \left\{ \dfrac{\emptyset(1 - S_w)}{B_{oi}} \right\}^{+1.042} \times \left(\dfrac{k_{\mu wi}}{\mu_{oi}} \right)^{+0.077} \times \left(S_w \right)^{-0.190} \times \left(\dfrac{P_i}{P_a} \right)^{-0.216}$

1. Connect 16.7% on oil in place scale *A* with mobility factor 0.103 on scale *G* and find intercept *b* on scale *B*.
2. Connect point *b* on scale *B* with water saturation 35% on scale *F* and find intercept *c* on scale *C*.
3. Connect point *c* on scale *C* with pressure drop ratio 2.48 on scale *E* and find the estimated recovery factor @554 barrels per acre-foot at intercept with scale *D*.

Figure 2.20 Nomograph for estimated recovery factor by water drive (in sandstone) (courtesy of SPE of AIME).

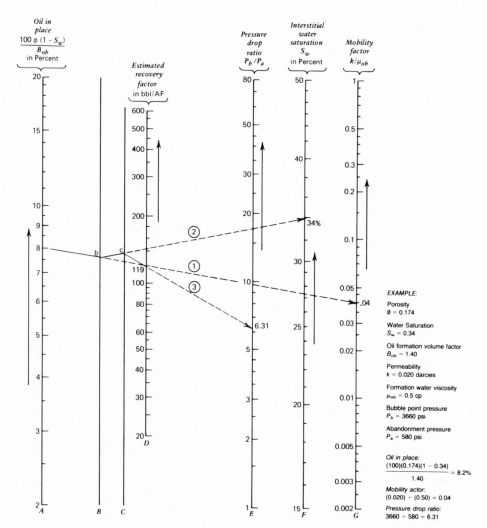

Based on: $BAF = 3244 \times \left\{ \dfrac{\emptyset(1 - S_w)}{B_{ob}} \right\}^{+0.161} \times \left(\dfrac{k}{\mu_{ob}} \right)^{+0.098} \times \left(S_w \right)^{+0.372} \times \left(\dfrac{P_b}{P_a} \right)^{+0.174}$

Oil in place
$\dfrac{100\,\phi\,(1 - S_w)}{B_{ob}}$
in Percent

Estimated recovery factor in bbl/AF

Pressure drop ratio P_b/P_a

Interstitial water saturation S_w *in Percent*

Mobility factor k/μ_{ob}

EXAMPLE:

Porosity
$\emptyset = 0.174$

Water Saturation
$S_w = 0.34$

Oil formation volume factor
$B_{ob} = 1.40$

Permeability
$k = 0.020$ darcies

Formation water viscosity
$\mu_{ob} = 0.5$ cp

Bubble point pressure
$P_b = 3660$ psi

Abandonment pressure
$P_a = 580$ psi

Oil in place:
$\dfrac{(100)(0.174)(1 - 0.34)}{1.40} = 8.2\%$

Mobility actor:
$(0.020) \div (0.50) = 0.04$

Pressure drop ratio:
$3660 \div 580 = 6.31$

1. Connect 8.2% on oil in place scale *A* with mobility factor 0.04 on scale *G* and find intercept *b* on scale *B*.
2. Connect point *b* on scale *B* with water saturation 34% on scale *F* and find intercept *c* on scale *C*.
3. Connect point *c* on scale *C* with pressure drop ratio 6.31 on scale *E* and find estimated recovery factor @119 barrels per acre-foot at intercept with scale *D*.

Figure 2.21 Nomograph for estimated recovery factor by solution-gas drive (below bubble point) (courtesy of SPE of AIME).

related as shown in Eq. 2.11 where the initial gas formation volume factor is given by:

$$B_{gi} = \frac{p_s T_i z_i}{p_i T_s z_s} \quad \text{ft}^3/\text{scf} \tag{2.22}$$

p_i and p_s = pressure at reservoir and standard conditions, psia

T_i and T_s = temperature at reservoir and standard conditions, °R

z_i *and* z_s = gas deviation factor at reservoir and base conditions, dimensionless

If B_{gi} is in res bbl/scf, Eq. 2.11 becomes

$$G = 7,758Ah\phi \, (1 - S_{wi}) \frac{1}{B_{gi}} \tag{2.23}$$

$$B_{gi} = \frac{p_s T_i z_i}{5.615 \, p_i T_s z_s} \quad \text{res bbl/scf} \tag{2.24}$$

At any subsequent reservoir pressure, the standard cubic feet of gas in place is given by:

$$G_x = 43,560Ah\phi \, (1 - S_w) \frac{1}{B_g} \tag{2.25}$$

$$B_g = \frac{p_s T z}{p T_s z_s} \tag{2.26}$$

If standard conditions are assumed to be 14.7 psia and 60°F and $z_s \simeq 1$, then Eq. 2.26 becomes

$$B_g = \frac{14.7}{p} \left(\frac{460 + T}{460 + 60} \right) z = 0.0283 \, (460 + T) \frac{z}{p} \tag{2.27}$$

The gas deviation factor should be handled properly because the omission of this factor in gas reserve calculations may introduce errors as large as 30 percent. The value of gas deviation factor used in Eq. 2.27 may be obtained from Fig. 2.22, if properties of the natural gas are known. Values of B_g and the reciprocal gas formation volume factor $(1/B_g)$ for different temperatures and pressures and for gases of specific gravities between 0.6 and 1.0 may be obtained from Figs. 2.23 to 2.27.

Example 2.4. Estimate gas in place in a reservoir with an areal extent of 2550 acres, average thickness of 50 ft, average porosity of 20 percent, connate water

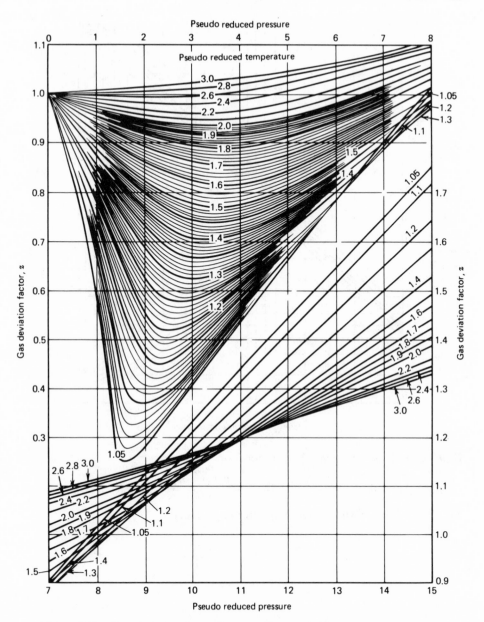

Figure 2.22   Gas deviation factor for natural gases (after Standing and Katz).

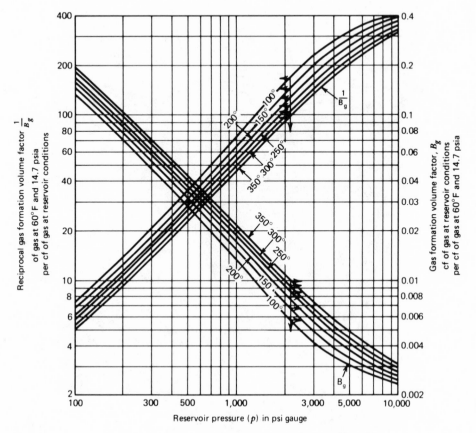

Figure 2.23   Gas-formation volume factor $B_g = \dfrac{14.7}{p_{\text{psig}} + 14.7} \dfrac{460 + T}{460 + 60} z$ and reciprocal gas-formation volume factor $\dfrac{1}{B_g} = \dfrac{p_{\text{psig}} + 14.7}{14.7} \dfrac{460 + 60}{460 + T} \dfrac{1}{z}$ vs. pressure, psig, and temperature, °F. Gas gravity 0.6 (air 1.0) (after Arps, 1962).

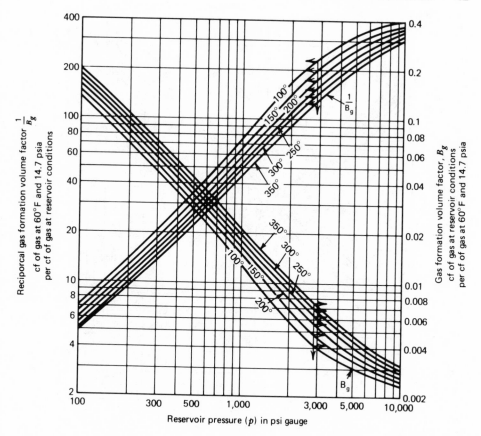

Figure 2.24   Gas-formation volume factor $B_g = \dfrac{14.7}{p_{psig} + 14.7} \dfrac{460 + T}{460 + 60} z$ and reciprocal gas-formation volume factor $\dfrac{1}{B_g} = \dfrac{p_{psig} + 14.7}{14.7} \dfrac{460 + 60}{460 + T} \dfrac{1}{z}$ vs. pressure, psig, and temperature, °F. Gas gravity 0.7 (air 1.0) (after Arps, 1962).

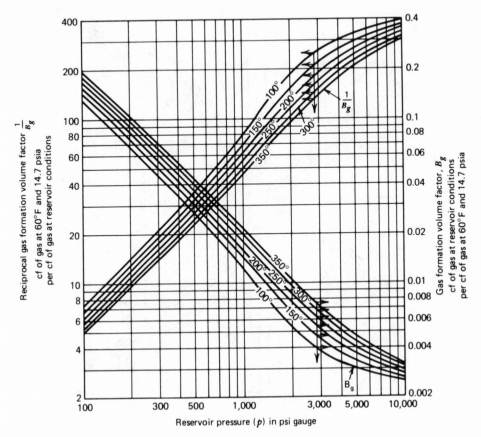

Figure 2.25   Gas-formation volume factor $B_g = \dfrac{14.7}{p_{\text{psig}} + 14.7}\dfrac{460 + T}{460 + 60}\, z$ and reciprocal

gas-formation volume factor $\dfrac{1}{B_g} = \dfrac{p_{\text{psig}} + 14.7}{14.7}\dfrac{460 + 60}{460 + T}\dfrac{1}{z}$ vs. pressure, psig, and tem-

perature, °F. Gas gravity 0.8 (air 1.0) (after Arps, 1962).

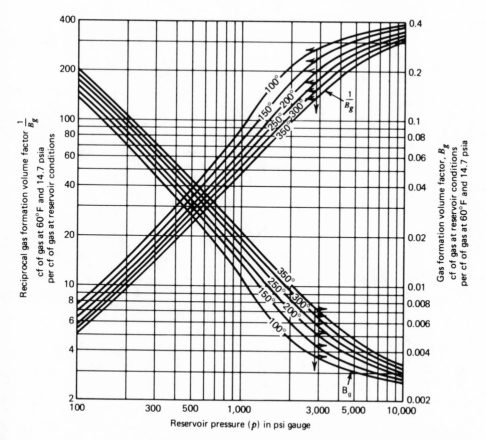

Figure 2.26   Gas-formation volume factor $B_g = \dfrac{14.7}{p_{\mathrm{psig}} + 14.7} \dfrac{460 + T}{460 + 60} z$ and reciprocal gas-formation volume factor $\dfrac{1}{B_g} = \dfrac{p_{\mathrm{psig}} + 14.7}{14.7} \dfrac{460 + 60}{460 + T} \dfrac{1}{z}$ vs. pressure, psig, and temperature, °F. Gas gravity 0.9 (air 1.0) (after Arps, 1962).

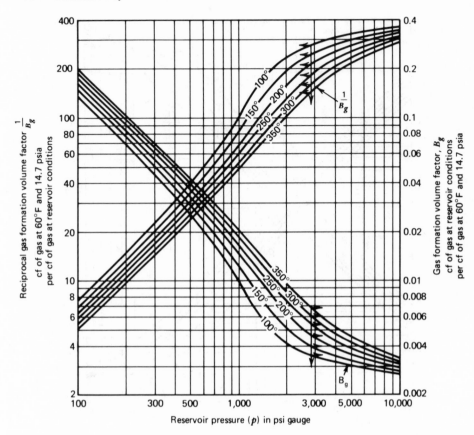

Figure 2.27   Gas-formation volume factor $B_g = \dfrac{14.7}{p_{\text{psig}} + 14.7} \dfrac{460 + T}{460 + 60} z$ and reciprocal

gas-formation volume factor $\dfrac{1}{B_g} = \dfrac{p_{\text{psig}} + 14.7}{14.7} \dfrac{460 + 60}{460 + T} \dfrac{1}{z}$ vs. pressure, psig, and tem-

perature, °F. Gas gravity 1.0 (air 1.0) (after Arps, 1962).

saturation of 20 percent, reservoir temperature of 186°F, initial reservoir pressure of 2651 psia, and recovery factor of 0.85. The composition of the gas is shown below.

Solution

Component (1)	Mole Fraction y_i (2)	Molecular Weight M_i (3)	Critical Temperature °R T_{ci} (4)	Critical Pressure psia p_{ci} (5)	y_iM_i (2) × (3) (6)	y_iT_{ci} (2) × (4) (7)	y_ip_{ci} (2) × (5) (8)
Methane	0.8602	16.04	343	668	13.798	295.0	574.6
Ethane	0.0770	30.07	550	708	2.315	42.4	54.5
Propane	0.0426	44.10	666	616	1.879	28.4	26.2
Isobutane	0.0057	58.12	735	529	0.331	4.2	3.0
Normal butane	0.0087	58.12	765	551	0.506	6.7	4.8
Isopentane	0.0011	72.15	829	490	0.079	0.9	0.5
Normal pentane	0.0014	72.15	845	489	0.101	1.2	0.7
Hexanes	0.0033	86.18	913	437	0.284	3.0	1.4
Total	1.0000				19.293	381.8	665.7

We then find the following:
Apparent molecular weight of gas,

$$M_a = \Sigma y_iM_i = 19.3 \text{ lbm/lb mol}$$

Specific gravity of gas,

$$\gamma_g = \frac{\text{molecular weight of gas}}{\text{molecular weight of air}} = \frac{M_a}{M_{air}} = \frac{19.3}{28.96} = 0.67$$

Pseudocritical temperature,

$$T_{pc} = \Sigma y_iT_{ci} = 381.8°R$$

Pseudocritical pressure,

$$P_{pc} = \Sigma y_ip_{ci} = 665.7 \text{ psia}$$

At a temperature of 186°F and pressure of 2651 psia:
Pseudoreduced temperature,

$$T_{pr} = \frac{T}{T_{pc}} = \frac{186 + 460}{381.8} = \frac{646°R}{381.8°R} = 1.69$$

Pseudoreduced pressure,

$$P_{pr} = \frac{P}{P_{pc}} = \frac{2{,}651 \text{ psia}}{665.7 \text{ psia}} = 3.98$$

From Fig. 2.22, $z = 0.855$. Equation 2.27 gives the gas formation volume factor as

$$B_g = 0.0283 \,(186 + 460)\,\frac{0.855}{2{,}651}$$

$$= 0.0059 \text{ ft}^3/\text{scf}$$

Interpolating between Figs. 2.23 and 2.24, we obtain

$$B_g = 0.006 \text{ ft}^3/\text{scf}$$

Using Eq. 2.11, we find that the recoverable gas reserves are

$$G = \frac{(43{,}560)(2{,}550)(50)(0.2)(1 - 0.20)(0.85)}{0.0059}$$

$$= 1{,}280{,}221 \text{ MMscf}^*$$

The volumetric equation (Eq. 2.11) is useful in reserve work for estimating gas in place at any stage of depletion. During the development period, before reservoir limits have been accurately defined, it is convenient to calculate gas in place per acre-foot of bulk reservoir rock. Multiplication of this unit figure by the best available estimate of bulk reservoir volume then gives gas in place for the lease, tract, or reservoir rock under consideration. Later in the life of the reservoir, when the reservoir volume is defined and performance data are available, volumetric calculations provide valuable checks on gas in place estimates obtained from material balance or other methods.

For natural gas reservoirs under volumetric control (no water influx or water production), the cumulative gas produced (G_p) at any pressure is the difference between the volumetric estimates of gas in place at the initial and subsequent pressure conditions. Thus, for volumetric reservoirs:

$$G_p = 43{,}560 A h \phi \,(1 - S_{wi}) \left(\frac{1}{B_{gi}} - \frac{1}{B_g}\right) \qquad (2.28)$$

If the gas formation volume factor B_{ga} at the assumed abandonment pressure (say 100 psia with and 500 psia without compressor facilities) is substituted for B_g, Eq. 2.28 gives G_p at abandonment or the recoverable gas in place at original conditions.

*The MM designation signifies one million; M is one thousand.

The recovery factor for a gas reservoir is primarily a function of the abandonment pressure and permeability. Lowering the abandonment pressure increases the recoverable gas. The abandonment pressure used depends on the price of gas, productivity indices of the wells, size of the field, its location with respect to the market, and the type of market. If the market is a transmission pipeline, the operating pressure of the line may be a controlling factor in the abandonment pressure for small fields; but for large fields, installation of compressor plants may be economically feasible, thus lowering the abandonment pressure substantially below the operating pressure of the pipeline serving the area. Some gas pipeline companies use an abandonment pressure of 100 psi/1000 ft of depth.

Water drive gas reservoirs usually have lower recovery factors than closed gas reservoirs because of the high abandonment pressure due to water encroachment into the wells. The reservoir permeability is also a primary factor governing the recovery from a closed gas reservoir. Higher permeabilities result in high flow rates for a given pressure drop. Therefore, when all other factors are the same, the abandonment pressure is lower for a high permeability reservoir. The recovery factor is high when the sand is uniform and homogeneous, the permeability of the sand is high at low gas saturation, the percentage of the gas-containing portion of the reservoir originally underlain by water is relatively small, the beds are relatively steep, and the amount of structural closure above the gas–water contact is large.

For closed gas reservoirs, the principal factor governing recovery factor is the abandonment pressure. If the abandonment pressure is known, a recovery factor can be calculated. Expressed in percent of initial gas in place, the recovery factor is

$$E_R = \frac{100 \, (B_{ga} - B_{gi})}{B_{ga}} = 100 \left(1 - \frac{B_{gi}}{B_{ga}}\right) = 100 \left(1 - \frac{p_a z_i}{p_i z_a}\right) \qquad (2.29)$$

For water drive reservoirs it is

$$E_R = \frac{100 \, (S_{gi} \, B_{ga} - S_{ga} \, B_{gi})}{S_{gi} \, B_{ga}} \qquad (2.30)$$

where

S_{gi} = initial gas saturation, fraction of initial pore space

S_{ga} = abandonment gas saturation, fraction of initial pore space

B_{gi} = initial gas formation volume factor, ft^3/scf or res bbl/scf

B_{ga} = abandonment gas formation volume factor, ft^3/scf or res bbl/scf

p_a = abandonment pressure, psia

z_a = gas deviation factor at abandonment

For strong water drives where residual gas is trapped at high pressures, E_R may be from 50 to 60 percent compared to 70 to 80 percent for partial water drives and 80 to 90 percent for volumetric reservoirs.

Table 2.5 provides values for residual gas saturation after water flood in core plugs, as measured by Geffen. These may be used in Eq. 2.30 as an approximation for S_{ga}.

Example 2.5. A proposed gas well is being evaluated. Well spacing is 640 acres, and it appears that the entire 640 acres attributed to this well is productive. Geological estimates indicate 30 ft of net effective pay, 15 percent porosity, and 30 percent interstitial water saturation. The pressure is 3000 psia and the reservoir temperature is 150°F. The abandonment pressure is estimated to be 500 psia. The gas has a specific gravity of 0.60. Base temperature and pressure are 60°F and 14.65 psia, respectively. An estimate of the gas reserves is required.

Solution
The first step is the calculation of B_{gi}, which requires pseudocritical T and p, pseudoreduced T and p, and then z, or read directly from Fig. 2.23. From Fig. 2.23, the pseudocritical pressure and temperature for a 0.6 gravity gas are 668 psia and 385°R, respectively.

$$\text{Pseudoreduced pressure} = \frac{3000 \text{ psia}}{668 \text{ psia}} = 4.5$$

$$\text{Pseudoreduced temperature} = \frac{(150 + 460)°R}{385°R} = 1.6$$

TABLE 2.5
Residual Gas Saturation After Water Flood
as Measured on Core Plugs

Porous Material	Formation	S_{gr} (%)
Unconsolidated sand		16
Slightly consolidated sand (synthetic)		21
Synthetic consolidated materials	Selas Porcelain	17
	Norton Alundum	24
Consolidated sandstones	Wilcox	25
	Frio	30–38
	Nellie Bly	30–36
	Frontier	31–34
	Springer	33
	Torpedo	34–37
	Tensleep	40–50
Limestone	Canyon Reef	50

Source: Geffen et al.

From Fig. 2.22, z_i is found to be 0.83. Using Eq. 2.26, we get

$$B_{gi} = \frac{(14.65)(150 + 460)(0.83)}{(3000)(60 + 460)(1.0)} = 0.004755 \text{ ft}^3/\text{scf}$$

Or, from Fig. 2.23,

$$B_{gi} = 0.0048 \text{ ft}^3/\text{scf}$$

The second step is to calculate the recovery factor, E_R. Abandonment pressure being 500 psia, the pseudoreduced pressure = 500/668 = 0.75. Using this value together with the pseudoreduced temperature of 1.6 in Fig. 2.22, we find z_a to be 0.94, and from Fig. 2.23, $1/B_{ga}$ is 0.032 ft^3/scf. Hence, from Eq. 2.29:

$$E_R = 100 \left(1 - \frac{0.0048}{0.032}\right) = 100 \left[1 - \frac{(500)(0.83)}{(3000)(0.94)}\right] = 85\%$$

The third step is to calculate reserves in scf/acre-ft:

$$G \text{ (scf/acre-ft)} = \frac{(43,560)(0.15)(1 - 0.30)(0.85)}{0.0048}$$

$$= 809,944 \text{ scf/acre-ft}$$

The final step is to multiply the scf/acre-ft figure by the net acre-feet; hence, the estimated gas reserves are:

$$G = (809,944 \text{ scf/acre-ft}) \times (640 \text{ acres})(30 \text{ ft})$$

$$= 15.6 \text{ billion scf of gas}$$

2.5   MATERIAL BALANCE ESTIMATES

When production of the hydrocarbons begins, the material balance method may be used to calculate the amount of oil or gas in place in a reservoir. This method is based on the premise that the pore volume of a reservoir remains constant or changes in a predictable manner with the reservoir pressure when oil, gas, and/or water are produced. This makes it possible to equate the expansion of the reservoir fluids upon pressure drop to the reservoir voidage caused by the withdrawal of oil, gas, and water minus the water influx. The use of the material balance method necessitates the collection of a lot of data, and the reserve estimates obtained is highly dependent upon the quality of the data.

Successful application of the material balance method requires an accurate history of the average pressure of the reservoir, as well as reliable oil-, gas-, and water-production data, and PVT data on the reservoir fluids. Generally, from 5 to

10 percent of the oil or gas originally in place must be withdrawn before significant results can be expected. Without very accurate performance and PVT data the results from such a computation may be quite erratic, especially when a water drive is present or the size of the free gas cap is unknown.

For oil reservoirs being produced by solution gas drive and/or water drive, the material balance may be expressed, in general, as:

(Oil zone expansion) + (water influx)

= (cumulative oil production) + (cumulative water production)

Oil zone expansion relates to the expansion of hydrocarbons within the original oil zone. This would be dominated by the release of solution gas. The water influx term represents the inflow of surrounding water into the reservoir.

The material balance equation in its most general form can be written as:

$$N = \frac{N_p \left[B_t + 0.1781 B_g (R_p - R_{si}) \right] - (W_e - W_p)}{B_{oi} \left\{ m \dfrac{B_g}{B_{gi}} + \dfrac{B_t}{B_{oi}} - (m + 1) \left[1 - \dfrac{\Delta p(C_f + S_w C_w)}{1 - S_w} \right] \right\}} \tag{2.31}$$

where

N = original oil in place, STB

N_p = cumulative oil produced at time t and pressure p, STB

B_o = oil formation volume factor at any time t, res bbl/STB

B_{oi} = initial oil formation volume factor, res bbl/STB

R_s = solution gas–oil ratio at any time t, scf/STB

R_p = cumulative gas–oil ratio, scf/STB

$$R_p = G_p/N_p$$

G_p = cumulative gas produced at any time t, scf

R_{si} = initial solution gas–oil ratio, scf/STB

B_g = gas formation volume factor, ft^3/scf

B_{gi} = initial gas formation volume factor, ft^3/scf

B_t = two-phase or total oil formation volume factor representing the volume in barrels occupied by one stock tank barrel of oil and its initial dissolved gas content at reservoir temperature T and any reservoir pressure p, res bbl/STB

$$B_t = B_o + (R_{si} - R_s) B_g \quad \text{and} \quad B_{ti} = B_{oi}$$

TABLE 2.6
Classification of Material-Balance Equations

Reservoir Type	Material-Balance Equation	Unknowns
Oil reservoir with gas cap and active water drive	$$N = \frac{N_p[B_t + 0.1781B_g(R_p - R_{si})] - (W_e - W_p)}{mB_{oi}\left(\dfrac{B_g}{B_{gi}} - 1\right) + (B_t - B_{oi})}$$	N, W_e, m
Oil reservoir with gas cap; no active water drive ($W_e = 0$)	$$N = \frac{N_p[B_t + 0.1781B_g(R_p - R_{si})] + W_p}{mB_{oi}\left(\dfrac{B_g}{B_{gi}} - 1\right) + (B_t - B_{oi})}$$	N, m
Inititally undersaturated oil reservoir with active water drive ($m = 0$): 1. Above bubble point	$$N = \frac{\left[N_p(1 + \Delta pC_o) - \dfrac{W_e - W_p}{B_{oi}}\right](1 - S_w)}{\Delta p[C_o + C_f - S_w(C_o - C_w)]}$$	N, W_e
2. Below bubble point	$$N = \frac{N_p[B_t + 0.1781B_g(R_p - R_{si})] - (W_e - W_p)}{B_t - B_{oi}}$$	N, W_e
Initially undersaturated oil reservoir; no active water drive ($m = 0$), ($W_e = 0$): 1. Above bubble point	$$N = \frac{\left[N_p(1 + \Delta pC_o) + \dfrac{W_p}{B_{oi}}\right](1 - S_w)}{\Delta p[C_o + C_f - S_w(C_o - C_w)]}$$	N
2. Below bubble point	$$N = \frac{N_p[B_t + 0.1781B_g(R_p - R_{si})] + W_p}{B_t - B_{oi}}$$	N
Gas reservoir with active water drive	$$G = \frac{G_pB_g - 5.615(W_e - W_p)}{B_g - B_{gi}}$$	G, W_e
Gas reservoir; no active water drive ($W_e = 0$)	$$G = \frac{G_pB_g + 5.615W_p}{B_g - B_{gi}}$$	G

Source: Arps 1962.

m = ratio of initial reservoir free gas volume to initial reservoir oil volume

$$m = \frac{GB_{gi}}{5.615\, NB_{oi}}$$

W_e = cumulative water influx, bbl

W_p = cumulative water produced, bbl

C_o = compressibility factor for reservoir oil, psi^{-1}

C_f = compressibility factor for reservoir rock, psi^{-1}

C_w = compressibility factor for interstitial water, psi^{-1}

S_w = interstitial water saturation, fraction of pore space

p = reservoir pressure, psia

The different forms the material balance equation may take are summarized in Table 2.6.

When a free-gas cap is present, the reservoir rock compressibility factor C_f and the interstitial water compressibility C_w may be neglected. Equation 2.31 may be simplified to Eq. 2.32, if there is an active water drive, and to Eq. 2.33, if there is no active water drive ($W_e = 0$).

For initially undersaturated reservoirs there is no initial free-gas cap ($m = 0$). If the reservoir is *above* the bubble-point pressure, no free gas is present ($R_p - R_{si} = 0$), while $B_t = B_{oi} + \Delta p C_o$. Thus, Eq. 2.31 reduces to Eq. 2.34 with active water drive, and Eq. 2.36 without active water drive. If the reservoir is *below* the bubble-point pressure, Eq. 2.32 reduces to Eq. 2.35 for active water drive case, and Eq. 2.33 becomes Eq. 2.37 if there is no active water drive.

The general material balance equation reduces to the gas reservoir material balance equation if N and N_p are set equal to zero. If an active water drive exists, Eq. 2.38 is obtained. If there is no active water drive, Eq. 2.39 applies.

REFERENCES

American Petroleum Institute: "A Statistical Study of Recovery Efficiency," *Bulletin D14*, API October, 1967.

Arps, J. J.: "Estimation of Primary Oil and Gas Reserves," Chapter 37 of *Petroleum Production Handbook*, edited by T. C. Frick and R. W. Taylor, McGraw-Hill, New York, 1962.

Arps, J. J.: "Reasons for Differences in Recovery Efficiency," SPE Reprint Series No. 3, 1970, pp. 49–54.

Geffen, T. M., D. R. Parrish, G. W. Haynes, and R. A. Morse: "Efficiency of Gas Displacement from Porous Media by Liquid Flooding," *Trans. AIME*, Vol. 195, 1952, pp. 29–38.

Hughes, R. V.: *Oil Property Valuation*, Robert E. Kreiger, Huntington, New York, 1978.

Meyer, R. F.: "The Volumetric Method for Petroleum Resource Estimation," *Mathematical Geology*, Vol. 10, No. 5, 1978, pp. 501–518.

Smith, C. R.: *Mechanics of Secondary Oil Recovery*, Robert E. Kreiger, Huntington, New York, 1975.

Standing, M. D. and D. L. Katz: "Density of Natural Gases," *Trans. AIME*, Vol. 146, 1942, pp. 140–149.

Wahl, W. L., L. D. Mullins, and E. B. Elfrink: "Estimation of Ultimate Recovery from Solution Gas-Drive Reservoirs," SPE Reprint Series No. 3, 1970, pp. 34–40.

3

PRODUCTION DECLINE CURVES

3.1 INTRODUCTION

Forecasting future production is the most important part of the economic analysis of exploration and production expenditures. Frequently this may be the weakest link in our analysis, for it may be based on little if any actual production performance. Analysis of production decline curves represents a useful tool for forecasting future production during capacity production from wells, leases, or reservoirs. The basis of this procedure is that factors which have affected production in the past will continue to do so in the future.

Decline curves can be characterized by three factors: (1) initial production rate, or the rate at some particular time, (2) curvature of the decline, and (3) rate of decline. These factors are a complex function of numerous parameters within the reservoir, well bore, and surface handling facilities. Formation parameters of porosity, permeability, thickness, fluid saturations, fluid viscosities, relative permeability, reservoir size, well spacing, compressibility, producing mechanism, and fracturing will all contribute to the character of the decline curve. Well bore conditions such as hole diameter, formation damage, lifting mechanism, solution gas, free gas, fluid level, completion interval, and mechanical conditions will have their effect on the decline curve too. The factors that directly affect the decline in production rate are: (1) reduction in average reservoir pressure, (2) increases in the field water cut in water drive fields, (3) increases in the field gas−oil ratio in depletion type fields, and (4) reduction in the liquid levels in gravity drainage fields. All comments apply to gas as well as oil wells.

A production record of an abandoned well and the known causes of changes in production rate are shown in Fig. 3.1. The projection of such a production decline curve into the future can be quite puzzling. Plotting of the average production rate of many wells in the reservoir with respect to time may iron out many irregularities.

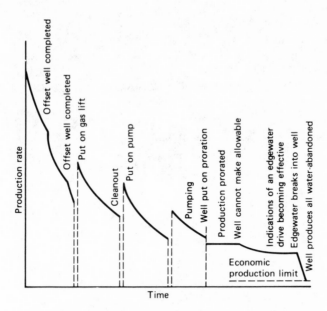

Figure 3.1 An idealized producti~~o~~n rate-time curve of an abandoned oil well (after Hughes).

Certain conditions must prevail before we can analyze a production decline curve with any degree of reliability. The production must have been stable over the period being analyzed. That is, a flowing well must have been produced with constant choke size or constant wellhead pressure, and a pumping well must have been pumped off or produced with constant fluid level. These indicate that the well must have been produced at capacity under a given set of conditions. The production decline observed should truly reflect reservoir productivity and not be the result of external causes, such as a change in production conditions, well damage, production controls, or equipment failure.

Stable reservoir conditions must also prevail in order to extrapolate decline curves with any degree of reliability. This condition will normally be met as long as the producing mechanism is not altered. However, when action is taken to improve the recovery of oil or gas, such as infill drilling, fluid injection, thermal operations, fracturing, and acidizing, decline curve analysis can be used to estimate the performance of the well or reservoir in the absence of the change and compare it to the actual performance with the change. Such a comparison enables us to determine the technical and economic success of our efforts.

Production decline curve analysis is used in the evaluation of new investments and the audit of previous expenditures. Associated with it is the sizing of equipment and facilities such as pipelines, plants, and treating facilities. Also associated with the economic analysis is the determination of reserves for a well, lease, or field. This is an independent method of reserve estimation, the result of which can be compared with volumetric or material balance estimates.

3.2 ECONOMIC LIMIT

The end point of the production decline curve is commonly called the economic limit. The economic limit rate is the production rate that will just meet the direct operating expenses of a well. In determining this economic limit it is advisable to analyze the expenditure charged against a well, and determine how much would actually be saved if the well were abandoned. Certain expenses may have to be continued if other wells on the lease are kept in operation. Overhead should be included only when abandonment contributes to a reduction in the overhead.

The economic limit can be written algebraically as

$$\text{Economic limit} = \frac{\text{direct operating costs}}{(\text{revenue} - \text{royalty})/\text{barrel}} \tag{3.1}$$

Thus, reduction in direct operating costs and increase in crude oil or natural gas price increases the amount of economically recoverable oil or gas, while increase in direct operating costs and reduction in crude oil or natural gas price causes a reduction in the economically recoverable oil or gas.

Example 3.1. Determine the economic-limit rate for a well using the following data:

Crude oil price per barrel	$30.00
Severance tax	5%
Ad valorem tax	3%
Royalty	12.5%
Estimated direct operating cost at economic limit	$2,800/month

Solution

$$\text{Net income per barrel} = \tfrac{7}{8}(1 - 0.05)(1 - 0.03)(\$30.00)$$

$$= \$24.19$$

$$\text{Economic limit} = \frac{\$2800/\text{month}}{(\$24.19/\text{bbl})(30.4\ \text{days/month})}$$

$$= 3.81\ \text{gross bbl/day} = 116\ \text{gross bbl/month}$$

3.3 CLASSIFICATION OF DECLINE CURVES

The production rate of wells, or groups of wells, generally declines with time. An empirical formula can sometimes be found that fits the observed data so well that it seems rather safe to use the formula to estimate a future relationship. The formulas relating time, production rate, and cumulative production are usually derived by first plotting the observed data in such a way that a straight-line relationship results. Some predictions can be made graphically by simply extrapolating the straight-line plot, or by the use of the mathematical formulas.

In most cases the production declines at a decreasing rate, that is, dq/dt decreases with time. Figure 3.2 shows an ideal curve. The $t = 0$ point can be chosen arbitrarily. q is the oil or gas production rate and t is time. The area under the curve between the times t_1 and t_2 is a measure of the cumulative production during this time period since

$$Q = \int_{t_1}^{t_2} q \, dt \tag{3.2}$$

There are three commonly recognized types of decline curves. Each of these has a separate mathematical form that is related to the second factor characterizing a decline curve, that is, the curvature. These types are referred to as:

1. Constant-percentage decline.

2. Harmonic decline.

3. Hyperbolic decline.

Each type of decline curve has a different curvature, as can be seen in Fig. 3.3. This figure depicts the characteristic shape of each type of rate versus time curve and rate versus cumulative curve on coordinate, semi-log, and log-log graph paper. For constant-percentage decline, rate versus time is a straight line on semi-log paper and rate versus cumulative is a straight line on coordinate paper. Rate versus cumulative is a straight line on semi-log paper for harmonic decline. All others have some curvature. Log-log plots of rate versus time for harmonic and rate versus cumulative for hyperbolic declines curves can be straightened out by using shifting techniques.

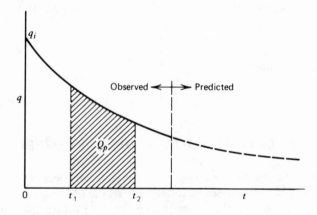

Figure 3.2 Graph of production rate versus time.

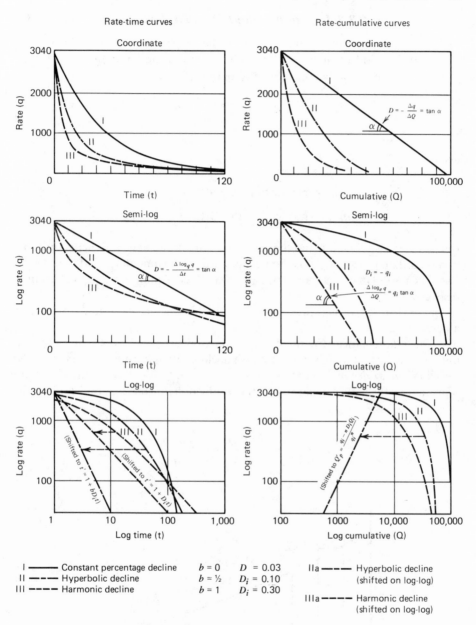

Figure 3.3 Three types of production decline curves.

3.3.1   Nominal and Effective Decline

The *effective decline rate* per unit time D' is the drop in production rate from q_t to q_{t+1} over a period of time equal to unity (1 month or 1 year) divided by the production rate at the beginning of the period (Fig. 3.4) or

$$D' = \frac{q_t - q_{t+1}}{q_t} = 1 - \frac{q_{t+1}}{q_t} \qquad (3.3)$$

where

$$q_t = \text{production rate at time } t$$

$$q_{t+1} = \text{production rate 1 time unit later}$$

Being a stepwise function and therefore in better agreement with actual production recording practices, D' is the decline rate more commonly used in practice. The time period may be 1 month or 1 year for effective monthly or annual decline, respectively.

From Eq. 3.3, D' is expressed as a fraction; in practice it is often expressed as a percentage.

The mathematical treatment of production decline curves is greatly simplified if the instantaneous or continuous decline rate is introduced. The *nominal (or continuous) decline rate D*, is defined as the negative slope of the curvature representing the natural logarithm of the production rate q versus time t (Fig. 3.5), or

$$D = - \frac{d(\ln q)}{dt} = - \frac{dq/dt}{q} \qquad (3.4)$$

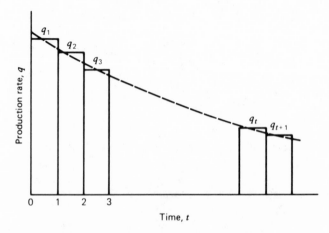

Figure 3.4   Production decline curve.

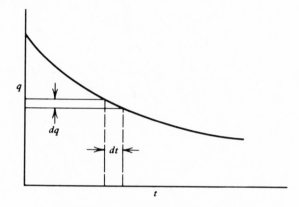

Figure 3.5 Production decline curve.

The second part of Eq. 3.4 shows that D can be visualized as the change in relative rate of production, dq/q, per unit of time. The minus sign has been introduced since dq and dt have opposite signs, and it is convenient to have D always positive.

Nominal decline, being a continuous function, is used mainly to facilitate the derivation of the various mathematical relationships. The decline rate in general changes with time except for the constant-percentage decline in which D is a constant. The relationship between D' and D will be derived later for the different production decline curves.

3.3.2 Constant Percentage Decline

A plot of production rate versus time is generally curved, but the plot of production rate versus cumulative production on Cartesian coordinate paper sometimes indicates a straight-line trend as shown in Fig. 3.6. The equation for the straight line can be written as

$$q = q_i - \alpha \, Q_D \tag{3.5}$$

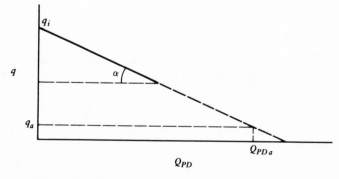

Figure 3.6 Rate versus cumulative production graph.

where

$$q_i = \text{production rate at the beginning of decline}$$
$$Q_D = \text{cumulative production when rate equals } q$$
$$\alpha = \text{slope of the straight line}$$

Other forms of Eq. 3.5 are

$$Q_D = \frac{q_i - q}{\alpha} \qquad \text{or} \qquad \alpha = \frac{q_i - q}{Q_D} \tag{3.6}$$

Differentiating Eq. 3.5 with respect to time yields

$$\frac{dq}{dt} = -\alpha \frac{d\,Q_D}{dt} \tag{3.7}$$

But

$$\frac{d\,Q_D}{dt} = q \tag{3.8}$$

Thus,

$$\frac{dq}{dt} = -\alpha q \tag{3.9}$$

From Eqs. 3.4 and 3.9 the continuous (nominal or instantaneous) decline rate is

$$D = -\frac{1}{q}\frac{dq}{dt} = \alpha \tag{3.10}$$

Thus, if q versus Q_D is a straight line, the nominal decline rate is equal to the slope of the straight line and is constant, hence the name *constant-percentage decline*.

The constant-percentage decline is the simplest, most conservative, and most widely used decline curve equation. This type of decline is the most frequently used for the following reasons:

1. Many wells and fields actually follow a constant-percentage decline over a great portion of their productive life, and then only deviate significantly toward the end of this period.

2. The mathematics of constant-percentage decline are much simpler and easier to use than the other two types of decline curves.

3. The divergence between a constant-percentage and the other types of decline occurs frequently quite a few years in the future. When this difference is discounted to the present time, it is not usually significant.

The differential equation that describes the constant-percentage decline is

$$D = -\frac{1}{q}\frac{dq}{dt} \quad (D = \text{constant}) \tag{3.11}$$

This states that the instantaneous or nominal decline rate is a constant percentage of the instantaneous production rate. The *rate-time* relation can be derived by integrating Eq. 3.11.

$$\int_{q_0}^{q} \frac{dq}{q} = -D \int_0^t dt \tag{3.12}$$

or

$$q = q_i e^{-Dt} \tag{3.13}$$

Since Eq. 3.13 is an exponential function, the constant-percentage decline is usually referred to as an *exponential decline*.

The *rate-cumulative* relationship may be obtained by integrating Eq. 3.13.

$$Q_D = \int_0^t q \, dt = q_i \int_0^t e^{-Dt} \, dt \tag{3.14}$$

or

$$Q_D = \frac{q_i (1 - e^{-Dt})}{D} = \frac{q_i - q}{D} \tag{3.15}$$

or

$$q = q_o - DQ_D \tag{3.16}$$

Equations 3.13 and 3.16 give rise to the basic plots used in the analysis of constant-percentage declines. Taking logarithms of Eq. 3.13 to base 10, we get

$$\log q = \log q_i - \frac{D}{2.303} t \tag{3.17}$$

where $2.303 = \ln 10$. A plot of $\log q$ versus t on Cartesian coordinate paper or q versus t on semi-log graph paper with q on the log scale will result in a straight line

(Fig. 3.7). The nominal decline rate is given by the slope of the plot on semi-log graph paper. A convenient formula for D is

$$D = \frac{2.303}{\Delta t/\text{cycle}} \tag{3.18}$$

where $\Delta t/\text{cycle}$ is the time difference between points that are a cycle apart on the q scale. Extrapolation of the straight line will yield future production rates until the economic limit, q_a is reached.

The second useful plot is based on Eq. 3.18. q versus Q_D plots as a straight line on Cartesian coordinates as shown in Fig. 3.6. The value of the nominal decline rate can be determined from the slope since

$$D = \frac{q_i - q}{Q_D} = \tan \alpha \tag{3.19}$$

The rate-cumulative plot is particularly useful for predicting production rates at future values of cumulative production. The reserves at any time can be determined by extrapolating the straight line to the economic-limit production, q_a or calculated from

$$Q_{Da} = \frac{q_i - q_a}{D} \tag{3.20}$$

The maximum amount of oil or gas producible regardless of economic considerations is obtained by extrapolating the straight line to $q = 0$ and is also given by q_i/D. This number is sometimes called the "movable oil" or "movable gas" as the case may be.

The dimension of the decline rate is 1/time. Since the product Dt is dimen-

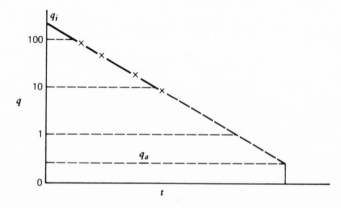

Figure 3.7   Graph of log q versus time.

sionless, the unit of D is the reciprocal of the unit of t used. If t is in months, D should be in 1/month, and so on.

From the definition of effective decline rate (Eq. 3.3),

$$q_1 = q_i(1 - D') \tag{3.21}$$

From Eq. 3.13, for 1 time unit

$$q_1 = q_i e^{-D} \tag{3.22}$$

From Eqs. 3.21 and 3.22

$$\frac{q_1}{q_i} = (1 - D') = e^{-D} = \frac{q_2}{q_1} = \frac{q_3}{q_2} = \frac{q_j}{q_{j-1}} = r \tag{3.23}$$

where r is the ratio of production rates of successive years. Thus,

$$D' = 1 - e^{-D} = 1 - r \tag{3.24}$$

and

$$D = -\ln(1 - D') = -\ln r \tag{3.25}$$

It is worth noting the relationship between the annual and monthly effective decline rate and between the annual and monthly nominal decline rate. If D'_m is the effective monthly decline rate, then from Eq. 3.23, the production rate at the end of the first month is $q_i(1 - D'_m)$; at the end of the second month it is $q_1(1 - D'_m) = q_i(1 - D'_m)^2$, and so on. Thus, at the end of 12 months the production rate is $q_i(1 - D'_m)^{12}$. But the production rate at the end of 12 months is also given by $q_i(1 - D'_a)$, where D'_a is the effective annual decline rate. Thus,

$$1 - D'_a = (1 - D'_m)^{12} \tag{3.26}$$

If D_m is the nominal monthly decline rate and D_a is the nominal annual decline rate

$$e^{-D_a} = (e^{-D_m})^{12} = e^{-12D_m} \tag{3.27}$$

or

$$D_a = 12D_m \tag{3.28}$$

Where producible reserves can be estimated from volumetric considerations and the initial and final production rates are known, the remaining life to abandonment time may be obtained by solving for time from Eqs. 3.13 and 3.15:

$$t_a = \frac{1}{D} \ln \left(\frac{q_i}{q_a} \right) \qquad (3.29)$$

or

$$t_a = \frac{Q_{Da}}{q_i - q_a} \ln \left(\frac{q_i}{q_a} \right) \qquad (3.30)$$

Under what conditions do we expect a constant-percentage decline to be applicable? Production above the bubble point follows this decline exactly. Production from solution gas drive reservoirs will approximate constant-percentage decline closely. Other producing mechanisms may also exhibit constant-percentage decline or reasonably approximate it; however, sufficient production data are needed to establish this fact.

Using Effective Annual Decline Rate

If $\bar{q}_t$ is the average annual production rate for year t, then the cumulative production for t years, Q_D, can be written as

$$Q_D = \bar{q}_1 + \bar{q}_2 + \bar{q}_3 + \cdots + \bar{q}_t \qquad (3.31)$$

For constant-percentage decline with an effective annual decline rate, D',

$$\bar{q}_t = \bar{q}_1 (1 - D')^{t-1} \qquad (3.32)$$

Substituting Eq. 3.32 in Eq. 3.31 yields

$$Q_D = \bar{q}_1 [1 + (1 - D') + (1 - D')^2 + \cdots + (1 - D')^{t-1}] \qquad (3.33)$$

Multiplying through Eq. 3.33 by $(1 - D')$ and subtracting the product from Eq. 3.31 gives

$$Q_D [1 - (1 - D')] = \bar{q}_1 [1 - (1 - D')^t] \qquad (3.34)$$

or

$$Q_D = \bar{q}_1 \left[\frac{1 - (1 - D')^t}{D'} \right] = \frac{\bar{q}_1 - \bar{q}_{t+1}}{D'} \qquad (3.35)$$

It should be noted that the average annual production rate for the first year, $\bar{q}_1$, is less than the instantaneous annual production rate at the beginning of the first year, q_i. There is a simple relationship between the two:

$$\bar{q}_1 = q_i \left(\frac{D'}{D} \right) \qquad (3.36)$$

Reserve to Production Ratio

Where performance data are available, the reserve-to-production (Q/q) ratio is a useful evaluation and screening tool. This ratio is strongly dependent on reservoir and fluid parameters as well as individual well and produced fluid conditions. For a given reservoir-producing mechanism and well conditions the true value of Q/q should be within a narrow range of values, which can usually be determined by analyzing other fields and/or reservoirs having similar characteristics.

Reserves to production ratio can be related mathematically to the remaining life of the producing unit being analyzed and the rate of annual production decline. If it is assumed that the production follows a constant-percentage decline until depletion, Eq. 3.35 can be written as:

$$Q_0 = \bar{q}_1 \left[\frac{1 - (1 - D')^t}{D'} \right] \tag{3.37}$$

where

Q_0 = remaining reserves at the end of previous year

$\bar{q}_0$ = previous year's production

$\bar{q}_1$ = current year's production

D' = effective annual decline rate

t = remaining life, years

By the definition of constant-percentage decline,

$$\bar{q}_1 = \bar{q}_0 (1 - D') \tag{3.38}$$

Substituting Eq. 3.38 in Eq. 3.37 and rearranging terms yields the following equation for ratio of year-end reserves to that year's production:

$$\frac{Q_0}{\bar{q}_0} = \left[\frac{1 - D'}{D'} \right] - \left[\frac{(1 - D')^{t+1}}{D'} \right] \tag{3.39}$$

Equation 3.39 is presented graphically in Fig. 3.8. This graph provides a quick method for determining reasonable values for Q/q, if the decline rate is known. If annual production rate is known, a reasonable range of reserves value can be determined. Even if the remaining life cannot be accurately forecast, a maximum value of Q/q and hence Q corresponding to an infinite remaining life can be determined.

Normally a producing unit will exhibit Q/q of between 2 and 10 during the middle two-thirds of its producing life. It is higher during the early development period and approaches 1.0 for the year before abandonment. A higher than normal value indicates either that the reserves are not fully developed or that they

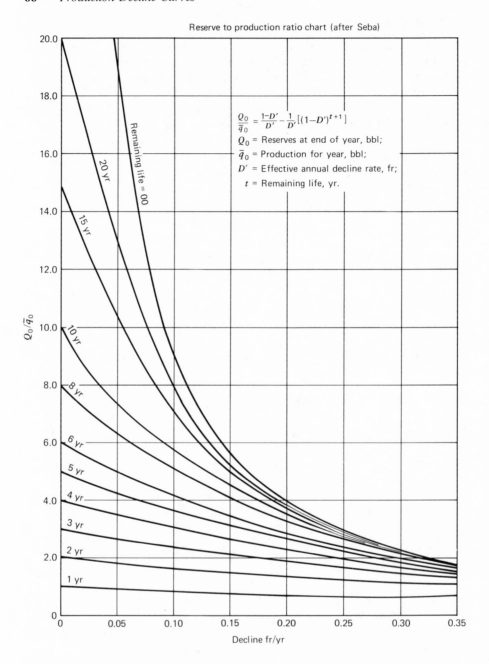

Reserve to production ratio chart (after Seba)

$$\frac{Q_0}{\bar{q}_0} = \frac{1-D'}{D'} - \frac{1}{D'}[(1-D')^{t+1}]$$

Q_0 = Reserves at end of year, bbl;
$\bar{q}_0$ = Production for year, bbl;
D' = Effective annual decline rate, fr;
t = Remaining life, yr.

Figure 3.8   Reserve to production ratio chart (after Seba).

are overstated. A high Q/q ratio occurs if there are significant reserves behind pipe awaiting future recompletion. Thus, a multipay or highly faulted field would exhibit a higher Q/q ratio than an unfaulted single layer reservoir. The high Q/q ratio also indicates that the reserve estimate is too high due to poor reservoir and/or geologic data or may indicate that the recovery efficiency is less than expected. Very tight reservoirs would also exhibit high Q/q ratios. Thus a high Q/q indicates that further evaluation is needed.

A low Q/q indicates that reserves may be understated or there may have been a recent change in production performance. High permeability reservoirs also tend to have lower Q/q than normal. Thus a low Q/q can also indicate that additional evaluation is needed.

Another graphical method of estimating constant-percentage decline is shown in Fig. 3.9. This approach allows a quick estimate of the five variables associated with constant-percentage decline—q_i, q_a, t, D', and Q_D. Although Figs. 3.8 and 3.9 are especially helpful for quick estimates and evaluations, they are not meant to replace the more precise mathematics of constant-percentage decline curve analysis.

Example 3.2. Using the following production data, estimate:

(a) The future production down to a rate of 50 bbl/day.
(b) Instantaneous (nominal or continuous) decline rate.
(c) Effective monthly and annual decline rates.
(d) Extra time necessary to obtain future production down to 50 bbl/day.

Production Data:

q (bbl/day)	N_P (bbl)	q (bbl/day)	N_P (bbl)
200	10×10^3	130	190×10^3
210	20×10^3	123	220×10^3
190	30×10^3	115	230×10^3
193	60×10^3	110	240×10^3
170	100×10^3	115	250×10^3
155	150×10^3		

Solution
A graph of q versus N_P is shown in Fig. 3.10 on Cartesian coordinates. A straight line is obtained indicating constant-percentage decline.

(a) From graph $N_p = 396,000$ bbl at $q = 50$ bbl/day. Future production = $396,000 - 250,000 = 146,000$ bbl.
(b) The nominal (instantaneous) decline rate is given by the slope of the straight line. Picking two points on the straight line

q (bbl/day)	N_p (bbl)
215	0
100	276,000

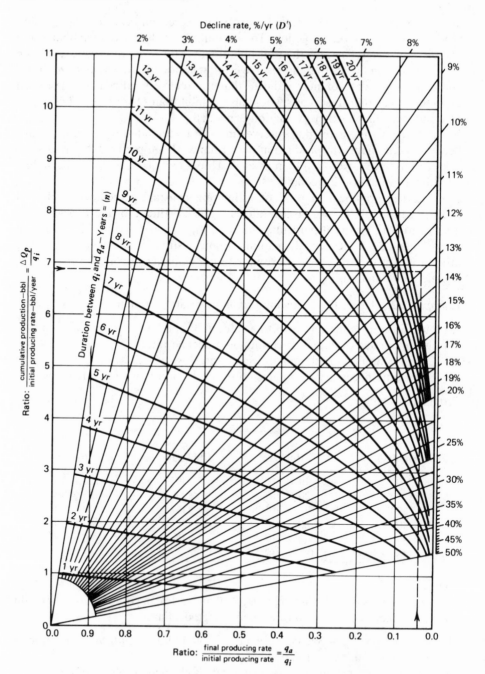

Figure 3.9 Exponential decline chart (after Schoemaker).

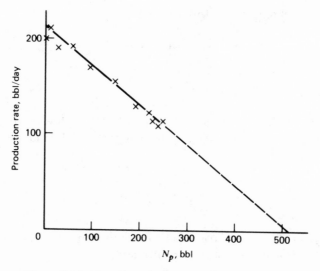

Figure 3.10 Rate-cumulative graph for example 3.2.

we find the nominal daily decline rate,

$$D_d = \frac{215 - 100}{276,000} = 0.000417 \text{ day}^{-1}$$

the nominal monthly decline rate,

$$D_m = 30.4 D_d = 0.0127 \text{ month}^{-1}$$

and the nominal yearly decline rate,

$$D_a = 12 D_m = 365 D_d = 0.152 \text{ year}^{-1}$$

(c) We also find the effective monthly decline rate,

$$D_m' = 1 - e^{-D_m} = 1.26\% \text{ month}^{-1}$$

and the effective annual decline rate,

$$D_a' = 1 - e^{-D_a} = 1 - (1 - D_m')^{12} = 14.1\% \text{ year}^{-1}$$

(d) The time to reach a production rate of 50 bbl/day or remaining life is obtained from Eq. 3.13 (starting at $t = 0$, $q = 115$ bbl/day):

$$50 = 115 e^{-0.152t}$$

$$t = \frac{\ln (50/115)}{-0.152} = 5.5 \text{ years}$$

Using Fig. 3.8 requires that we calculate

$$\frac{N_P}{q} = \frac{146,000}{(115)(365)} = 3.5$$

$$D'_a = 14.1\%$$

Remaining life from Fig. 3.8 is 4.5 years, which is slightly less than the value calculated using decline curve equations. Using Fig. 3.9 requires calculation of

$$\frac{q_a}{q_i} = \frac{50}{115} = 0.43$$

Figure 3.9 gives $D'_a = 15$ percent per year and remaining life of 5.2 years, which agree reasonably well with our calculations.

Example 3.3.   Consider a gas well with the following production history for the year 1982.

Date	Production Rate (MMscf/month)
1-1-82	1,000
2-1-82	962
3-1-82	926
4-1-82	890
5-1-82	860
6-1-82	825
7-1-82	795
8-1-82	765
9-1-82	735
10-1-82	710
11-1-82	680
12-1-82	656
1-1-83	631

(a) Plot these data on semi-log graph paper to investigate the type of decline.
(b) Calculate the reserves to be produced from 1-1-83 to the economic limit of 25 MMscf/month.
(c) When will the economic limit be reached?
(d) How much gas will be produced each year until the economic limit is reached?

Solution
(a) The plot of q versus t on semi-log graph paper (Fig. 3.11) indicates a straight-line trend; therefore, constant-percentage decline is assumed.
(b) The reserves at economic-limit production rate can be calculated from Eq. 3.20,

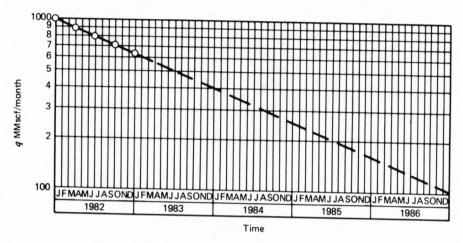

Figure 3.11   Example 3.3 constant percentage decline.

$$G_{PDa} = \frac{q_i - q_a}{D}$$

The nominal decline rate, D, can be determined from the rate-time equation or from the slope of rate-time plot on semi-log graph paper. Using two points on the straight line, we get $t = 0$, $q_i = 1000$ and $t = 12$, $q = 631$.

$$631 = 1000e^{-12D}$$

which gives

$$D = 0.0384 \text{ per month}$$

Or, from the slope, using Eq. 3.18,

$$D = \frac{2.303}{60 \text{ months/cycle}} = 0.0384 \text{ per month}$$

Thus,

$$G_{PDa} = \frac{1,000 - 25}{0.0384} = 25,391 \text{ MMscf}$$

(c) The life of the gas well is given by

$$25 = 1000e^{-0.0384t}$$

and

$$t = 96 \text{ months or } 8 \text{ years}$$

(d) The production each year is given by

$$G_{PD} = \frac{q_i - q}{D}$$

where

$$q_i = \text{rate at start of year}$$

$$q = \text{rate at end of year}$$

Year	q_i	q	G_{PD} (MMscf)
1983	631	398	6,068
1984	398	251	3,828
1985	251	158	2,422
1986	158	100	1,510
1987	100	63	964
1988	63	40	599
1989	40	25	391

3.3.3 Harmonic Decline

A graph of production rate versus cumulative may not show a straight-line trend on Cartesian coordinate paper. This graph sometimes shows a straight-line trend when replotted on semi-log graph paper (log q versus Q_D) as shown in Fig. 3.3. The equation for such a straight line is

$$\ln q = \log q_i - \alpha Q_D \tag{3.40}$$

That is,

$$Q_D = \frac{1}{\alpha} \ln \frac{q_i}{q} \tag{3.41}$$

or

$$q = q_i e^{-\alpha Q_D} \tag{3.42}$$

Differentiating Eq. 3.42 with respect to time gives us

$$\frac{dq}{dt} = -\alpha q_i e^{-\alpha Q_D} \frac{dQ_d}{dt} = -\alpha q^2 \tag{3.43}$$

From which

$$D = - \frac{1}{q} \frac{dq}{dt} = \alpha q \qquad (3.44)$$

and

$$D_i = \alpha q_i \qquad (3.45)$$

We can now eliminate α from Eqs. (3.44) and (3.45):

$$\frac{D_i}{q_i} = \frac{D}{q} \quad \text{or} \quad D = \frac{D_i}{q_i} q \qquad (3.46)$$

Equation 3.46 indicates that the nominal decline rate is not constant but decreases proportionally with the production rate. This is called a *harmonic* decline.

The rate-time relation can be obtained by integrating the basic equation

$$D = - \frac{dq/dt}{q} = \frac{D_i}{q_i} q \qquad (3.47)$$

$$- \int_{q_i}^{q} \frac{dq}{q^2} = \frac{D_0}{q_0} \int_{0}^{t} dt \qquad (3.48)$$

$$\frac{1}{q} = \frac{1}{q_i} + \frac{D_i}{q_i} t \qquad (3.49)$$

or

$$q = \frac{q_i}{1 + D_i t} \qquad (3.50)$$

The *cumulative-time* and rate-cumulative relationships can be obtained by integrating Eq. 3.50:

$$Q_D = \int_{0}^{t} q \, dt = q_0 \int_{0}^{t} \frac{dt}{1 + D_i t} \qquad (3.51)$$

that is,

$$Q_D = \frac{q_i}{D_i} \ln (1 + D_i t) \qquad (3.52)$$

or in terms of the rate of production

$$Q_D = \frac{q_i}{D_i} \ln \frac{q_i}{q} \tag{3.53}$$

The two basic plots for harmonic decline curve analysis are based on Eqs. 3.49 and 3.53. Equation 3.49 indicates that a plot of $1/q$ versus t on Cartesian coordinates will yield a straight line (Fig. 3.12). The intercept on the $1/q$ axis at $t = 0$ is $1/q_i$, and the slope of the line is D_i/q_i from which D_i may be directly determined.

Writing Eq. 3.53 as

$$\ln q = \ln q_i = \frac{D_i}{q_i} Q_D \tag{3.54}$$

or

$$\log q = \log q_i - \frac{D_i}{2.303 q_i} Q_D \tag{3.55}$$

it can be seen that a plot of q versus Q_D on semi-log graph paper will yield a straight line that can be extrapolated to the economic limit to give us the economically recoverable reserves (see Fig. 3.13). The reserves at abandonment are given by

$$Q_{Da} = \frac{q_i}{D_i} \ln \frac{q_i}{q_a} \tag{3.56}$$

Movable oil or gas is not defined for harmonic decline. The slope of the straight line is equal to $D_i/2.303 q_i$. This yields the same value for D_i as Fig. 3.12 provided that the value of q_i is known.

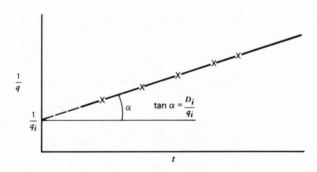

Figure 3.12 $1/q$ versus time graph (harmonic decline).

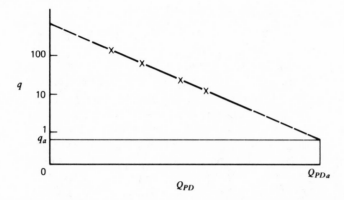

Figure 3.13 Log q versus cumulative production graph (harmonic decline).

The remaining life to abandonment time may be obtained as

$$t_a = \frac{q_i/q_a - 1}{D_i} \tag{3.57}$$

or

$$t_a = \frac{Q_{Da}}{q_i} \frac{(q_i/q_a) - 1}{\ln (q_i/q_a)} \tag{3.58}$$

The relationships between effective and nominal decline rates are

$$D_i' = \frac{D_i}{1 + D_i} \tag{3.59}$$

$$D_i = \frac{D_i'}{1 - D_i'} \tag{3.60}$$

There are several conditions under which harmonic decline has been observed in crude oil production. Production of high viscosity oil driven by encroaching edgewater is a good example. Because of the unfavorable mobility ratio, early breakthrough occurs and the bulk of oil production is obtained at high water cuts. If gross production (oil + water) is kept constant, then the increasing amount of water in the total fluid will cause the oil production to decline. It has also been noted that harmonic decline occurs under some thermal operations such as steam soak in thick gravity drainage reservoirs. Harmonic decline is not widely employed in reserves estimation.

3.3.4 Hyperbolic Decline

If the graph of log production rate versus time is curved, a straight-line relationship can still be obtained by adjusting and replotting the data on log-log graph paper. The process is known as *shifting a curve*. It amounts to the addition of a positive or negative constant to the variable to be plotted on the log scale. For decline curve analysis shifting is usually done on log-log graph paper, even though it can also be done on semi-log graph paper.

A replot of the data might be made in the form of log q versus log $(t + c)$ where c is an arbitrary constant. The amount of curve displacement, c, could be determined by trial and error, but less tedious methods are available.

The equation of the straight line obtained after shifting is

$$\log q = \log q_i - 1/b \ [\log (t + c) - \log c] \tag{3.61}$$

where b is the reciprocal of the slope (a positive constant), or

$$q = q_i \left(1 + \frac{t}{c}\right)^{-1/b} \tag{3.62}$$

Equation 3.62 shows that a plot of log q versus log $[1 + (t/c)]$ would also yield a straight line with a slope, $1/b$.

From Eq. 3.62,

$$\frac{dq}{dt} = -\frac{1}{bc} q_i \left(1 + \frac{t}{c}\right)^{-1/b-1} \tag{3.63}$$

and

$$D = -\frac{1}{q}\frac{dq}{dt} = \frac{1}{bc}\left(1 + \frac{t}{c}\right)^{-1} = \frac{1}{bc} q^b \tag{3.64}$$

From Eq. 3.64, when $t = 0$

$$D_i = \frac{1}{bc} \quad \text{or} \quad c = \frac{1}{bD_i} \tag{3.65}$$

Putting Eq. 3.65 in Eq. 3.64, we obtain

$$D = D_i (1 + bD_i t)^{-1} \quad \text{or} \quad \frac{1}{D} = \frac{1}{D_i} + bt \tag{3.66}$$

Equation 3.66 indicates a straight-line relationship between $1/D$ and t, which may sometimes be useful in determining D_i and b. The slope of the straight line is b and the intercept on the $1/D$ axis (at $t = 0$) is $1/D_i$.

The *rate-time* relationship is obtained by putting Eq. 3.65 in Eq. 3.62:

$$q = q_i (1 + bD_i t)^{-1/b} \tag{3.67}$$

Equation 3.67 may be written as

$$q^{-b} = q_i^{-b} (1 + bD_i t) \tag{3.68}$$

This indicates that a graph of q^{-b} versus t on Cartesian coordinate paper will yield a straight line with slope $bD_i q_i^{-b}$ and intercept of q_i^{-b} (at $t = 0$). A value of b is assumed and then checked by the linearity of q^{-b} versus t (see Fig. 3.14). The correct value of b will yield the best straight line.

Comparing Eqs. 3.66 and 3.67, we obtain

$$\left(\frac{q}{q_i}\right)^b = \frac{D}{D_i} \tag{3.69}$$

This shows that the hyperbolic decline includes both the constant-percentage and harmonic declines. From Eqs. 3.10, 3.46, and 3.69, $b = 0$ yields equal-percentage decline and $b = 1$ yields harmonic decline. Thus the limits of the hyperbolic decline constant are $0 \leq b \leq 1$.

The *rate-cumulative* relationship is obtained by integrating Eq. 3.67:

$$Q_D = q_i \int_0^t \frac{dt}{(1 + bD_i t)^{1/b}} \tag{3.70}$$

or

$$Q_D = \frac{q_i}{(1 - b)D_i} [1 - (1 + bD_i t)^{(b-1/b)}] = \frac{q_i}{(1 - b)D_i} \left[1 - \left(\frac{q}{q_i}\right)^{1-b} \right] \tag{3.71}$$

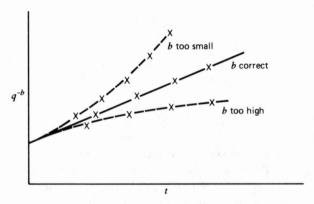

Figure 3.14 Graph of q^{-b} versus time (hyperbolic decline).

and

$$Q_D = \frac{q_i^b}{(1-b)D_i} [q_i^{(1-b)} - q^{(1-b)}] \tag{3.72}$$

The cumulative production down to the economic limit becomes

$$Q_{Da} = \frac{q_i}{(1-b)D_i} \left[1 - \left(\frac{q_a}{q_i} \right)^{(b-1/b)} \right] \tag{3.73}$$

The remaining time on decline is given by

$$t_a = \frac{1-b}{b} \frac{Q_{Da}}{q_i} \left(\frac{q_i}{q_a} \right)^{1-b} \frac{(q_i/q_a)^b - 1}{(q_i/q_a)^{1-b} - 1} \tag{3.74}$$

The movable oil or gas (at $q = 0$) is $q_i/(1-b)D_i$.

Under certain conditions production obtained by gravity drainage will follow hyperbolic decline with a hyperbolic decline constant $b = 1/2$. The rate-time relationship then becomes

$$q = \frac{q_i}{[1 + (D_i/2)t]^2} \tag{3.75}$$

and the rate-cumulative relationship

$$Q_D = \frac{2\sqrt{q_i}}{D_i} (\sqrt{q_i} - \sqrt{q}) \tag{3.76}$$

The remaining life to abandonment for this special case of hyperbolic decline ($b = 1/2$) is

$$t_a = \frac{2(\sqrt{q_i/q_a} - 1)}{D_i} \tag{3.77}$$

or

$$t_a = \frac{Q_{Da}}{q_i} \sqrt{q_i/q_a} \tag{3.78}$$

Gas wells usually produce at constant rates as prescribed by gas contracts. During this period the well pressure declines until it reaches a minimum level dictated by the line or compressor intake pressure. Thereafter, the well produces at a declining rate. If at this stage the square of the bottom hole flowing pressure is still much smaller than the square of the reservoir pressure, the decline is approximately hyperbolic with b equal to 1/2.

The effective decline rate and nominal decline rate for hyperbolic decline are related as follows:

$$D'_i = 1 - (1 + bD_i)^{-1/b} \tag{3.79}$$

$$D_i = \frac{1}{b}[(1 - D'_i)^{-b} - 1] \tag{3.80}$$

A curve fitting procedure based on reading three points from a smooth curve representing a set of data points is the most direct method of analyzing hyperbolic decline curves. The procedure is as follows:

1. Plot data as production rate versus time on semi-log graph paper and draw a smooth curve through them (Fig. 3.15).

2. Select two points 1 and 2 on the smooth curve giving (q_1,t_1) and (q_2,t_2). Points 1 and 2 are chosen arbitrarily on the smooth curve; the only restriction being that they should lie as closely as possible at the ends of the curve.

3. Calculate the third point 3 corresponding to (q_3,t_3). The value of q_3 is obtained from

$$q_3 = (q_1 \, q_2)^{0.5} \tag{3.81}$$

The corresponding value of t_3 is read from the smooth curve.

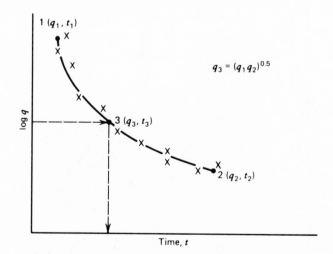

Figure 3.15 Curve fitting for hyperbolic decline.

4. Calculate the amount of shift:

$$c = \frac{t_1 t_2 - t_3^2}{t_1 + t_2 - 2t_3} \tag{3.82}$$

5. Shift data on log-log graph paper by adding $-c$ to the time values (Fig. 3.16). The numerical value of $c = 1/bD_i$.

6. Draw the q versus $(1 + bD_i t)$ curve that should be a straight line.

7. Use points read from the q versus $(1 + bD_i t)$ curve and the equation obtained by taking the logarithm of Eq. 3.67:

$$\log q = \log_i - \frac{1}{b} \log (1 + bD_i t) \tag{3.83}$$

to determine values of q_i, b, and D_i.

A graphical method for determining the value of b quickly is given in Figs. 3.17 and 3.18 any time q_i/q is less than 100. These figures can also be used for extrapolating decline curves to some future point. Outside the range of these figures the original equations from which these figures were drawn should be used. This is a quick estimate only and should not be used to replace the more precise approaches of earlier discussions.

To determine the value of the hyperbolic decline constant from Fig. 3.17, enter the abscissa (Q/tq_i) with values corresponding to the last data point on the

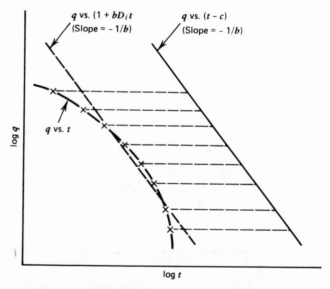

Figure 3.16   Shifting a curve on log-log graph paper.

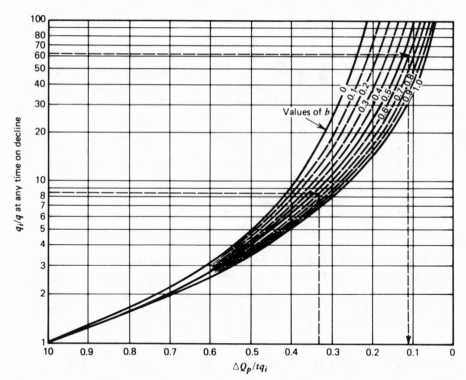

Figure 3.17 Relationship between production rate and cumulative production (after Gentry).

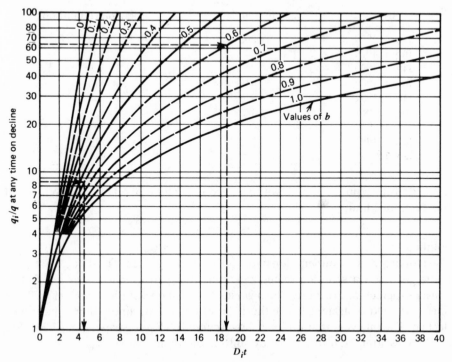

Figure 3.18 Relationship between production rate and time (after Gentry).

decline curve, and enter the ordinate (q_i/q) with the value of the ratio of initial production rate on the decline curve to that for the last data point. The hyperbolic decline constant is obtained by the intersection of these two values. The initial decline rate can be determined from Fig. 3.18 by entering the ordinate with the value of q_i/q used in Fig. 3.17 and moving right to the curve for the value of b determined from Fig.3.17. The initial decline rate, D_i, is then the value read on the abscissa divided by the time from q_i to q. These curves can be used for extrapolation by reversing the procedure, starting either with the terminal rate or time.

These graphs can also be used to analyze constant-percentage and harmonic decline curves since those two types are special cases of hyperbolic decline. The curves for $b = 0$ are for constant-percentage decline, and for harmonic decline, $b = 1$.

A recent paper by Fetkovich presents some insight into decline curve analysis. It demonstrates that decline curve analysis not only has a solid fundamental base but also provides a tool with more diagnostic power than has been suspected previously. The use of type curves for decline curve analysis is demonstrated with examples. Some of the type curves are shown in Figs. 3.19, 3.20, and 3.21.

Example 3.4. The following production data are available for a well:

Date	Daily Production Rate (bbl/day)	Cumulative Production (1000 bbl)
Jan. 1, 1979	1,000	0
July 1, 1979	840	167
Jan. 1, 1980	712	308
July 1, 1980	616	430
Jan. 1, 1981	536	535
July 1, 1981	472	627
Jan. 1, 1982	418	708
July 1, 1982	372	778
Jan. 1, 1983	336	844

Estimate future production down to an economic limit of 50 bbl/day. When will this economic limit be reached?

Solution
The plot of q versus t on semi-log paper is shown in Fig. 3.22. The data do not yield a straight line on semi-log paper and, thus, the performance does not follow constant-percentage decline. The q versus N_{PD} plot on semi-log paper (Fig. 3.23) does not yield a straight line either; therefore, the decline is not a harmonic decline. The rate versus time plot is shown curved on log-log paper (Fig. 3.24). We shall try to straighten this curve by shifting on log-log paper.

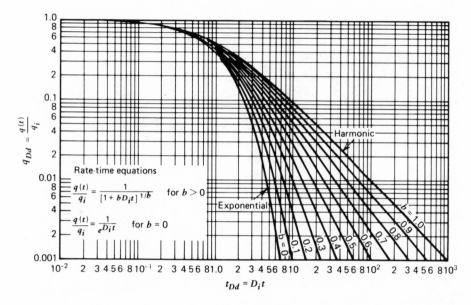

Figure 3.19 Type curves for Arps empirical rate-time decline equations, unit solution ($D_i = 1$) (after Fetkovich).

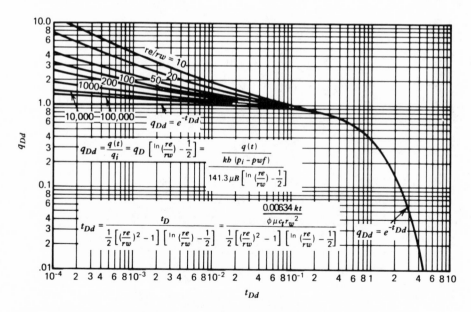

Figure 3.20 Dimensionless flow rate functions for plane radial system, infinite and finite outer boundary, constant pressure at inner boundary (after Fetkovich).

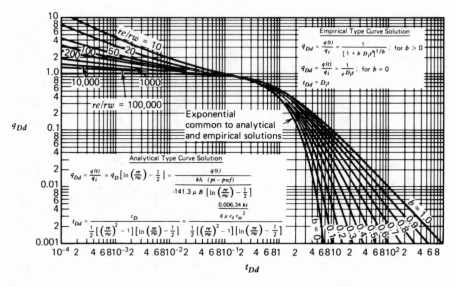

Figure 3.21 Composite of analytical and empirical type curves of Figures 3.19 and 3.20 (after Fetkovich).

From Fig. 3.22,

$$\text{Point 1: } q_1 = 940 \qquad t_1 = 0.2$$
$$\text{Point 2: } q_2 = 350 \qquad t_2 = 3.8$$
$$q_3 = (940 \times 350)^{0.5} = 574$$

From the curve, $t_3 = 1.75$

$$c = \frac{(0.2)(3.8) - (1.75)^2}{0.2 + 3.8 - 2(1.75)} = -4.61$$

q versus $(t + 4.61)$ and q versus $(1 + bD_i t)$ or q versus $(1 + 0.217t)$ straight lines are shown in Fig. 3.24. Picking two points on the q versus $(1 + 0.217t)$ line, we get

$$A: q = 1,000 \qquad (1 + 0.217t) = 1.0$$
$$B: q = 145 \qquad (1 + 0.217t) = 3.0$$

Using Eq. 3.83 gives us

$$3 = \log q_i - 0$$

$$2.161 = \log q_i - \frac{1}{b}(0.477)$$

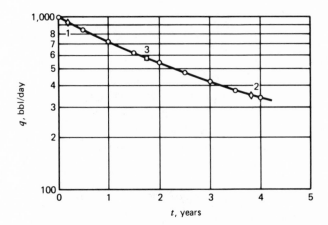

Figure 3.22 Rate-time plot for example 3.4.

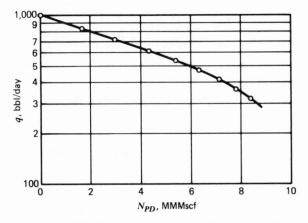

Figure 3.23 Rate-cumulative plot for example 3.4.

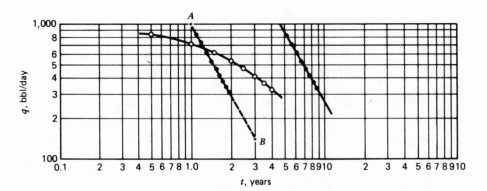

Figure 3.24 Shifting curve on log-log paper for example 3.4.

Thus,

$$b = 0.57$$

$$q_i = 1,000$$

$$D_i = 1/(4.61 \times 0.57) = 0.38/\text{year}$$
$$= 0.032/\text{month or } 0.001/\text{day}$$

Using Eq. 3.71, we find that the remaining reserves are

$$N_{PDa} = \frac{336}{(1 - 0.57)(0.001)}\left[1 - \left(\frac{50}{336}\right)^{0.43}\right]$$

$$= 437,000$$

From Eq. 3.74,

$$t_a = \frac{(1 - 0.57)}{0.57}\ \frac{(437,000)}{336}\left(\frac{336}{50}\right)^{0.43}\ \frac{(336/50)^{0.57} - 1}{(336/50)^{0.43} - 1}$$

$$= 3443 \text{ days} \qquad \text{or} \qquad 9.43 \text{ years}$$

The values of b and D_i may be obtained using the technique of Eq. 3.66:

Time	q	$-\Delta q$	q_{av}	$\dfrac{1}{D} = \dfrac{-q}{\Delta q/\Delta t}$	t_{av}
0	1,000				
0.5	840	160	920	2.88	0.25
1.0	712	128	776	3.03	0.75
1.5	616	96	664	3.46	1.25
2.0	536	80	576	3.60	1.75
2.5	472	64	504	3.94	2.25
3.0	418	54	445	4.12	2.75
3.5	372	46	395	4.29	3.25
4.0	336	36	354	4.92	3.75

The plot of $1/D$ versus t yields a straight line (Fig. 3.25). From the slope $b = 0.55$ and the intercept gives $D_i = 0.38$ per year. These agree quite well with values obtained earlier.

Let us now try to determine the value of b using Fig. 3.17.

$$N_{PD}/tq_i = 844,000/(4)(365)(1000) = 0.58$$

$$q_i/q = 1000/336 = 2.98$$

Figure 3.17 gives a value of $b \approx 0.5$, and from Fig. 3.18, $D_i t = 1.5$ or $D_i = 1.5/4$ or 0.38 per year.

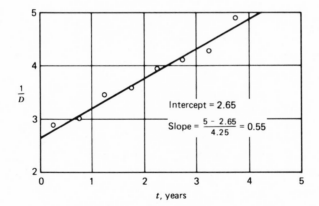

Figure 3.25 $1/D$ vs. time for example 3.4.

3.4 FRACTION OF RESERVES PRODUCED AT A RESTRICTED RATE

A common application of decline curves arises in the calculation of the schedule of production to be expected from a new well. In this case we must consider that the production rate from the well may be limited because of proration during the early years, or by physical items such as limited capacity of flow lines or transportation facilities. Later, the well will decline, but the decline rate must be based on analogy with other wells in the area that are already in decline. Although the greatest uncertainties are associated with predicting the time pattern in the later stages of production, these later stages are generally heavily discounted in arriving at a present value. Inaccuracies in this portion are relatively unimportant.

Figure 3.26 illustrates a common approximation for estimating the time pattern of production where the rate is restricted. Relationship for the fraction of reserves produced under restricted or allowable production can be derived from rate-cumulative relationships. For constant-percentage decline,

$$\frac{Q_r}{Q} = \frac{q_i - q_r}{q_i - q_a} \tag{3.84}$$

where q_r is allowable or has a restricted production rate. For harmonic decline,

$$\frac{Q_r}{Q} = \frac{\ln (q_i/q_r)}{\ln (q_i/q_a)} \tag{3.85}$$

For hyperbolic decline,

$$\frac{Q_r}{Q} = \frac{1 - (q_r/q_i)^{1-b}}{1 - (q_a/q_i)^{1-b}} \tag{3.86}$$

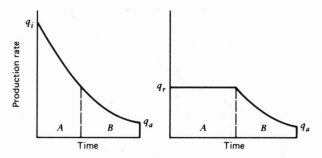

Figure 3.26 Estimating the effect of restricting maximum production rate.

Example 3.5. It has been determined from volumetric calculations that the ultimate recoverable reserves for a proposed well are 30 MMMscf of gas. By analogy with other wells in the area,

$$\text{Nominal decline rate} = 0.04 \text{ month}$$

$$\text{Allowable production rate} = 400 \text{ MMscf/month}$$

$$\text{Economic limit} = 30 \text{ MMscf/month}$$

What is the production by year for this well?

Solution
Since no production history exists for this well, exponential or constant-percentage decline is assumed.
From Eq. 3.15

$$G_{PD} = \frac{400 - 30}{0.04} = 9250 \text{ MMscf}$$

From Eq. 3.29

$$t_D = \frac{1}{0.04} \ln \left(\frac{400}{30} \right) = 65 \text{ months or 5.4 years}$$

Reserves during proration,

$$= (30,000 - 9250) \text{ MMscf} = 20,750 \text{ MMscf}$$

Time during restricted production,

$$t_r = \frac{20,750}{400} = 52 \text{ months or 4.3 years}$$

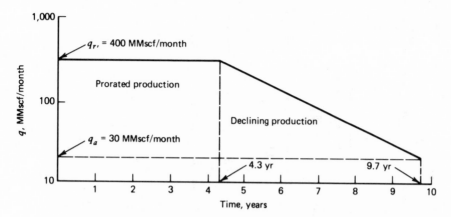

Figure 3.27 Production schedule for example 3.5.

We can now prepare our production forecast as shown in Fig. 3.27.

Production for the first four years = 400 × 12 = 4800 MMscf/year

Production for the fifth year can be divided into four months at constant production plus eight months at declining production.

For first four months: 4 × 400 = 1600 MMscf

At end of fifth year: $q = 400e^{-(0.04 \times 8)} = 290$ MMscf/month

For last eight months $G_{PD} = \dfrac{400 - 290}{0.04} = 2739$ MMscf

Total production for fifth year = 4339 MMscf

For subsequent years, production is given by $(q_i - q_{end})/D$. The results are summarized below:

Year	Production (MMscf/year)
1	4,800
2	4,800
3	4,800
4	4,800
5	4,339
6	2,765
7	1,710
8	1,050
9	650
10	325
Total	30,039

TABLE 3.1
Classification of Production Decline Curves

Decline Type	I. Constant-Percentage Decline	II. Hyperbolic Decline	III. Harmonic Decline
Basic Characteristic	Decline is constant $b = 0$	Decline is proportional to a fractional power (b) of the production rate $0 < b < 1$	Decline is proportional to production rate $b = 1$

I. Constant-Percentage Decline

$$D = K \cdot q^0 = -\frac{dq/dt}{q}$$

$$\int_0^t D\, dt = -\int_{q_i}^{q_t} \frac{dq}{q}$$

$$-Dt = \ln \frac{q_t}{q_i}$$

II. Hyperbolic Decline

$$D = K \cdot q^b = -\frac{dq/dt}{q}$$

For initial conditions:

$$K = \frac{D_i}{q_i^b}$$

$$\int_0^t \frac{D_i}{q_i^b} \cdot dt = -\int_{q_i}^{q_t} \frac{dq}{q^{b+1}}$$

$$\frac{bD_i t}{q_i^b} = q_t^{-b} - q_i^{-b}$$

III. Harmonic Decline

$$D = K \cdot q^1 = -\frac{dq/dt}{q}$$

For initial conditions:

$$K = \frac{D_i}{q_i}$$

$$\int_0^t \frac{D_i}{q_i} \cdot dt = -\int_{q_i}^{q_t} \frac{dq}{q^2}$$

$$\frac{D_i t}{q_i} = \frac{1}{q_t} - \frac{1}{q_i}$$

Rate-time relationship

$q_t = q_i \cdot e^{-Dt}$	$q_t = q_i(1 + bD_it)^{-1/b}$	$q_t = q_i(1 + D_it)^{-1}$
$Q_{PD} = \int_0^t q_t \cdot dt = \int_0^t q_i \cdot e^{-Dt} \cdot dt$	$Q_{PD} = \int_0^t q_t \cdot dt = \int_0^t (1 + nD_it)^{-1/b} \cdot dt$	$Q_{PD} = \int_0^t q_t \cdot dt = \int_0^t q_i(1 + D_it)^{-1} \cdot dt$
$Q_{PD} = \dfrac{q_t - q_i \cdot e^{-Dt}}{D}$	$Q_{PD} = \dfrac{q_i}{(b-1)D_i}[(1 + bD_it)^{(b-1)/b} - 1]$	$Q_{PD} = \dfrac{q_i}{D_i}[\ln(1 + D_it)]$
Substitute from rate-time equation:	Substitute from rate-time equation:	Substitute from rate-time equation:
$q_i \cdot e^{-Dt} = q_t$	$(1 + bD_it) = \left(\dfrac{q_i}{q_t}\right)^b$	$(1 + D_it) = \dfrac{q_i}{q_t}$
To find:	To find:	To find:

Rate-cumulative relationship

$Q_{PD} = \dfrac{q_i - q_t}{D}$	$Q_{PD} = \dfrac{q_i b}{(1-b)D_i}(q_i^{1-b} - q_t^{1-b})$	$Q_{PD} = \dfrac{q_i}{D_i}\ln\dfrac{q_i}{q_t}$

Symbols:

D = decline as a fraction of production rate
D_i = initial decline
q_i = initial production rate
t = time

q_t = production rate at time t
Q_{PD} = cumulative production at time t
K = constant
b = exponent

TABLE 3.2
Summary of Decline Equations

	Exponential (b = 0)	Harmonic (b = 1)	Hyperbolic (b = 1/2)	General Hyperbolic (b = b)
Decline factor	$D = -\ln(1 - D')$	$D_i = \dfrac{D_i'}{1 - D_i'}$	$D_i = 2\left(\dfrac{1}{\sqrt{1-D_i'}} - 1\right)$	$D_i = 1/b\left[\left(\dfrac{1}{1-D_i'}\right)^b - 1\right]$
Life, t_a	$t_a = \dfrac{\ln(q_i/q_a)}{D}$	$t_a = \dfrac{(q_i/q_a) - 1}{D_i}$	$t_a = \dfrac{2}{D_i}(\sqrt{q_i/q_a} - 1)$	$t_a = \dfrac{1}{bD_i}\left[\left(\dfrac{q_i}{q_a}\right)^b - 1\right]$
Rate at any time, t	$q = q_i e^{-Dt}$	$q = \dfrac{q_i}{1 + D_i t}$	$q = \dfrac{q_i}{\left(1 + \dfrac{D_i t}{2}\right)^2}$	$q = \dfrac{q_i}{(1 + bD_i t)^{1/b}}$
Cumulative production $Q_{PD} = f(t)$	$Q_{PD} = \dfrac{q_i}{D}(1 - e^{-Dt})$	$Q_{PD} = \dfrac{q_i}{D_i}\ln(1 + D_i t)$	$Q_{PD} = \dfrac{2q_i t}{2 + D_i t}$	$Q_{PD} = \dfrac{q_i}{D_i}\dfrac{1}{1-b}[1 - (1 + bD_i t)^{(b-1)/b}]$
Cumulative production $Q_{PD} = f(q)$	$Q_{PD} = \dfrac{q_i - q}{D}$	$Q_{PD} = \dfrac{q_i}{D_i}\ln\dfrac{q_i}{q}$	$Q_{PD} = \dfrac{2}{D_i}(q_i - \sqrt{q_i q})$	$Q_{PD} = \dfrac{q_i}{D_i}\dfrac{1}{1-b}\left[1 - \left(\dfrac{q}{q_i}\right)^{(1-b)}\right]$
Cumulative production $Q_{PDa} = f\left(t_a, \dfrac{q_i}{q_a}\right)$	$Q_{PDa} = \dfrac{q_i t_a(1 - q_a/q_i)}{\ln(q_i/q_a)}$	$Q_{PDa} = \dfrac{q_i t_a \ln(q_i/q_a)}{(q_i/q_a) - 1}$	$Q_{PDa} = \dfrac{q_i t_a}{\sqrt{q_i/q_a}}$	$Q_{PDa} = \dfrac{bq_i t_a}{1 - b}\left[\dfrac{\left(\dfrac{q_a}{q_i}\right)^b - \left(\dfrac{q_a}{q_i}\right)}{1 - \left(\dfrac{q_a}{q_i}\right)^b}\right]$

Symbols:

D = decline factor—exponential decline
D_i = initial decline factor—hyperbolic decline
D' = decline rate as a decimal
b = hyperbolic exponent

t = time in general
t_a = life of well
q_a = production rate at time, t_a
q = production rate at any time, t

q_i = initial production rate
Q_{PD} = cumulative production at time, t
Q_{PDa} = cumulative production at time, n
$\ln$ = natural logarithm

3.5 SUMMARY

Tables 3.1 and 3.2 summarize the development and the pertinent relationship for the three types of decline curves we have discussed. Decline-curve analysis is a useful tool for reserves estimation and production forecasts. Combined with the time value of money it can be used to simplify the economic analysis of exploration and producing projects. Decline curves also serve as diagnostic tools and may indicate the need for stimulation or remedial work.

REFERENCES

Arps, J. J.: "Analysis of Decline Curve," *Trans. AIME*, Vol. 160, 1945, p. 228.

Arps, J. J.: "Estimation of Primary Oil Reserves," *Trans. AIME*, Vol. 207, 1956, p. 182.

Arps, J. J.: "Estimation of Primary Oil and Gas Reserves," Chapter 37 of *Petroleum Production Handbook*, edited by T. C. Frick, McGraw-Hill, New York, 1962.

Campbell, R. A. and J. M. Campbell, Sr.: *Mineral Property Economics*, Vol. 3: Petroleum Property Evaluation, Campbell Petroleum Series, Norman, Oklahoma, 1978.

Fetkovich, M. J.: "Decline Curve Analysis Using Type Curves," *Journal of Petroleum Technology*, June 1980, p. 1065.

Gentry, R. W.: "Decline Curve Analysis," *Journal of Petroleum Technology*, January 1972, p. 38.

Hughes, R. V.: *Oil Property Valuation*, Robert E. Krieger, New York, 1978.

McCray, A. W.: *Petroleum Evaluations and Economic Decision*, Prentice-Hall, Englewood Cliffs, New Jersey, 1975.

Nind, T. E. W.: *Principles of Oil Well Production*, McGraw-Hill, New York, 1981.

Root, P. J.: "Curve Fitting," Chapter 5 of *Gas Lift Theory and Practices*, edited by K. E. Brown, Prentice-Hall, Englewood Cliffs, New Jersey, 1967.

Seba, R. D.: "Estimation of Economically Recoverable Oil from Decline Curve Analysis," Lecture Notes, Stanford University 1976.

Schoemaker, R. P.: *World Oil 1967*, October, p. 123.

4

CASH FLOW

4.1 INTRODUCTION

Projects and ventures cost money, the spending of which usually has to be justified in terms of receiving a return or making a profit. Economic evaluation consists of comparing the costs of the resources and effort needed to finance these ventures with the value of the expected benefits. In order to make this comparison as many factors as possible are measured in the same units, those of money or cash. The money unit used throughout this book is the dollar ($).

Any project (or venture) involves a certain set of payments (cash movements). Some of these payments are made by the Company to the project in order to run it. These are called *cash-out items*. Other payments are made to the Company as a result of the project. These are *cash-in items*. The net result of the payments for any particular period, say, a year, will either be a cash-out or cash-in, which may be indicated by negative or positive sign, respectively. A series of such annual cash deficit/surplus figures for successive years represents the *cash flow* of the project.

Net cash flow is the basis for all economic decisions. It is formed from the year-by-year sums of projected investments, income, and expenses. It may be developed on either a before tax basis or an after tax basis. Items included in a complete after tax cash flow analysis are both *cash items* (monies actually received or paid) and *noncash items* (bookkeeping items usually required for federal income tax calculations). Some of the cash items in oil and gas economic evaluation are working interest revenue, state and local taxes, operating costs, overhead, capital investments, bonus and leasehold costs, property sales, federal income taxes, capital gains taxes, investment tax credit, and windfall profit tax. Noncash items include depreciation, depletion, amortization, deferred deductions, capitalized investments, expensed investments, and book value.

4.2 TERMS AND CONCEPTS

To set the stage for work directed toward the economic evaluation of a project, it is necessary to become conversant with the terms and concepts related to the financial aspects of a business.

Revenues are the funds received by the Company during the period under consideration.

Operating income includes all of the money taken in by the Company from the sale of crude oil, natural gas, natural gas liquids, and refined products.

Costs and expenses are the costs of doing business that must be paid from the revenues received by the Company.

Taxes include federal, foreign, and other income taxes paid directly by the Company. In addition to income taxes, the Company's tax bill includes just about every type of levy governments have been able to devise. Leading the list are severance taxes imposed on crude oil and natural gas production and local property taxes on Company installations.

Intangible drilling and development costs (IDC) refer to certain expenses incurred by the Company in drilling to develop new crude oil and natural gas reserves. Primarily, these costs include money spent for site preparation, road building, fuel, repairs, contract services and other intangible items, and also the cost of dry holes. Generally the costs of these items that have no salvage value (as distinguished from steel pipe, pumping equipment, and other equipment of tangible nature) are deducted as incurred.

Some of the examples of IDC are all amounts of money paid for labor, fuel, repairs, hauling, and supplies, used in the following ways:

a. In the drilling, perforating, and cleaning of wells.

b. In preparation for the drilling of wells such as clearing of ground, draining, road making, surveying and geological works.

c. In the construction of derricks, tanks, pipelines, and other physical structures as are necessary for the drilling of wells and the production of oil and gas.

These costs include the cost of installation of tangible equipment placed in the well itself although the equipment placed in the well is to be capitalized and depreciated.

Equipment costs (tangible) refer to expenditures for items such as materials placed in the well, casing, tubing, tanks, wellhead, flowlines, etc. These costs are charged against income through depreciation. Equipment costs qualify for investment tax credit.

Depreciation recognizes the decline in value of machines and plants due to wear and tear of normal use and to obsolescence. Depreciation allowance is a method of capital recovery. It is the amortization of capitalized investment so as to spread cost over a period of time for tax or other bookkeeping purposes.

Cost depletion refers to the reduction in value of the investment in producing oil and gas acreage. Primarily, this is money spent for mineral rights, including the purchase price or lease bonuses and leasehold costs paid to land owners. Cost depletion allowance is designed to amortize these costs as provided by federal tax laws.

Percentage depletion allowance is a specified percentage of gross income after royalties from the sale of minerals removed from the mineral property during the tax year. Only small oil and gas producers may take percentage depletion allowance.

Amortization is similar to depreciation, whereby the costs of certain assets (e.g., patents, copyrights, goodwill, pollution control devices) are charged against income over a period of years.

Book value refers to the current values in the capital accounts not yet recovered by depreciation or depletion.

Leasehold refers to costs related to acquisition of mineral interests from land-owners. They include (a) acreage acquisition costs (landowner's bonus), (b) geophysical and geological costs, (c) legal expenses, and (d) title recording costs.

Mineral interests is the sum of all rights to oil and gas or to solid minerals in place. It is the owner of the minerals, not the surface owner, who has the right to grant an oil and gas or solid mineral lease. Also, it is the owner of the mineral rights who is entitled to any lease bonus and delay rentals that may be payable in connection with the lease.

Royalty interest is a right to oil and gas or minerals in place that entitles its owner to a specific fraction, in kind or in value, of the total production from the property, free of development and operating expenses. Traditionally this fraction ranges from 1/8 to 3/16, but has been increasing in the recent times.

A royalty interest may be created by assignment, as when the owner of the mineral rights conveys an interest therein, reserving to himself all operating rights and burdens. It may be reserved by the owner of the minerals when he grants an oil and gas lease or mining claim giving to the grantee the exclusive rights and burdens of development.

Working interest is an interest in minerals in place that bears the development and operating costs of the property. The working interest is created or carved out of the mineral interest by the granting of an oil and gas or mineral lease or mining claim. The lessee usually is under no obligation to develop the property or pay delay rentals, but there is an agreement that the lease will expire if the property is not developed or rentals are not paid, or if a specified amount of exploratory or assessment work is not done.

Overriding royalty is an additional royalty created from the working interest and having the same terms as this interest. It is said to be *carved out* if the owner of the working interest assigns a right to a fractional share of production free and clear of development and operating costs. It is said to be *retained* if the lessee assigns the working interest and retains a fractional share of production free of development and operating costs. The latter is more common and usually results from the activities of land broker who obtains leases for the sole purpose of ultimately assigning them to a third party for development purposes. This form of royalty may also become part of a *farmout* agreement.

Carried interest is a common type of sharing arrangement in the oil and gas industry. It involves a carried party who owns a working interest in mineral property but does not share in the working interest revenue until a certain amount of the money has been recovered, and a carrying party who pays for the drilling, development, and operations of a property for a period of time.

4.3   BASIC ELEMENTS OF THE NET CASH FLOW

The cash flow for any unit of time differs from the accounting net profit for that period because of the way the deductions and capitalized investments are treated. Depreciation and depletion allowances, amortization, and deferred deductions must be added back to the accounting profit because they are not "out of pocket" expenditures of the operation. They are merely bookkeeping adjustments for use of equipment and machinery or for depletable resources. They are permissible deductions for income tax purposes, however, and therefore affect the cash flow indirectly. The capitalized investment is an actual expenditure. It is not a permissible deduction item for the period in which it occurred, but it is deducted in computing the net cash flow.

Because sales dollars come in intermittently and some of the items of operating costs may be irregular, it is necessary to have a reservoir of *working capital* to be drawn on as needed to facilitate and insure smooth operation. The net flow of working capital in and out of the operation is zero. Working capital is not an allowable deduction. From experience each company determines the proper amount of working capital it needs for smooth operation without tying up too much capital.

A typical flow of funds for a company is illustrated in Fig. 4.1. The excess of sales over operating costs is called gross profit. From this stream depreciation, depletion, amortization, and deferred deductions are diverted leaving taxable income. The taxable income is subject to state and federal income taxes. The net profit remaining after these taxes are deducted, together with depreciation, depletion, amortization, and deferred deductions is termed cash flow, and this is shown in the cash flow as flowing into the company's bank.

Data used for computation of profitability criteria are developed in what is usually called a *net cash flow table*, an example of which is given in Table 4.1. The columns are explained in the pages following Table 4.1.

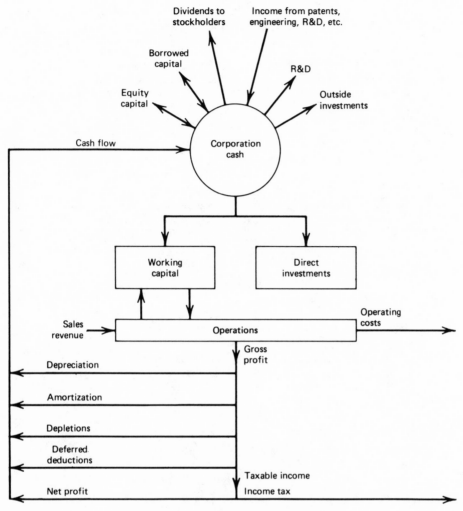

Figure 4.1 Corporate cash flow diagram.

4.3.1 Future Production Rates

Any profitability study must be based on the most reliable data available, and these data must be obtained before an economic study can be started. The importance of reliable production-rate estimates cannot be overemphasized. The reliability of the results of the profitability analysis is never better than the basic data used in the calculations. Methods of estimating oil and gas reserves and future production rates have been described in Chapters 2 and 3.

The development of a field depends on the geology of the field and interpretation of petrophysical data, in addition to the drilling program. Data provided by

TABLE 4.1
The Net Cash Flow Table

Year	Gross Production Oil	Gas	WI Production Oil	Gas	Revenue Oil	Gas	Total	Capital Investment	Number of Producing Wells
1									
2	(1)	(2)	(3)	(4)	(5)	(6)	(7)	(8)	(9)
↓									
Totals									

Year	Expenses Operation and Maintenance	Ad Valorem Tax	Total	Net Revenue After Expense	State and Federal Income Tax	Net Cash Flow
1						
2	(10)	(11)	(12)	(13)	(14)	(15)
↓						
Totals						

Year	Depreciation	Net Revenue After Expense and Depreciation	Depletion (Statutory) Gross	Net	Allowable	Net Revenue After Expense, Depreciation, and Depletion	State and Federal Income Tax
1							
2	(16)	(17)	(18)	(19)	(20)	(21)	(22)
↓							
Totals							

Depletion (Cost)
(20A)

Cash Flow Detail

Columns (1) and (2)   This is normally the total future oil and gas production forecasted from the property in question. This value is determined from your analysis. It is normally determined on an annual basis.

Column (3)   *Working Interest Production—Oil*
Column (1) times the Working Interest Fraction. (If you are evaluating a lease in which you have a 100% ownership, your Working Interest Fraction is (1 − Oil Royalty Fraction). If you are evaluating a lease in which you only have a portion of the ownership, your Working Interest Fraction is your portion times (1 − Oil Royalty Fraction).

Column (4)   *Working Interest Production—Gas*
Column (2) times the Working Interest Fraction; same as for the oil *but* use the Gas Royalty Fraction.

Column (5)   *Revenue—Oil*
Column (3) times Oil Value.

Column (6)   *Revenue—Gas*
                    Column (4) times Gas Value.

Column (7)   *Revenue—Total*
                    Column (5) plus (6).

Column (8)   *Capital Investment*
                    Amount and timing determined with help from other Engineering Groups.
                    It is normally broken into "tangible" and "intangible" for use in tax
                    calculations.

Column (9)   *Number of Producing Wells*
                    This column is used only if operation and maintenance is to be forecasted
                    on a cost per well-year basis. The number of producing wells is determined
                    from your analysis of future performance and is directly related to column
                    (1) and (2), Gross Production.

Column (10)   *Operation and Maintenance Expense*
                    Your analysis of future cost. It is column (9) times cost per well-year if your
                    forecast is on a well-year basis.

Column (11)   *Ad Valorem Tax*
                    Normally column (3) times the forecasted tax per Working Interest barrel.

Column (12)   *Total Expense*
                    Column (10) plus (11).

Column (13)   *Net Revenue After Expense*
                    Column (7) minus (12).

Column (14)   *State & Federal Income Tax*
                    From column (22) of Income Tax Detail.

Column (15)   *Net Cash Flow*
                    Column (13) minus column (8) and column (14).

Column (16)   *Depreciation*
                    Normally determined on a unit of production basis. The annual value is the
                    ratio of the remaining amount to be depreciated to the remaining reserves
                    times the production for the year. Other basis may be used.

Column (17)   *Net revenue after expense and depreciation*
                    Column (13) of Cash Flow Detail minus intangible portion of Column (8)
                    (Cash Flow Detail) minus column (16).

Column (18)   *Depletion—gross*
                    20% of column (7) (Cash Flow Detail). This percentage depletion may not
                    be taken by large producers.

Column (19)   *Depletion—net*
                    50% of Column (17). This is the limit for percentage depletion.

Column (20)   *Depletion—allowable*
                    The smaller of (18) or (19). *Note:* If cost depletion is applicable, Allowable
                    Depletion is the larger of Cost and Statutory. Statutory Depletion is *always*
                    the smaller of Gross and Net.

Column (21)   *Net Revenue after expense, depreciation, and depletion*
                    Column (17) minus (20)

Column (22)   *State and federal income tax*
                    Column (21) times effective tax rate.

Column (20A)   *Depletion cost*
                    The portion of the purchase price and/or bonus that is allocated to lease-
                    hold is used for the base for cost depletion. Normally determined on a unit
                    of production basis. The annual value is the ratio of the remaining amount
                    to be depleted to the remaining reserves times the production for the year.

reservoir engineering studies should be taken into account. These include estimated oil and gas reserves, reservoir production mechanism, PVT characteristics
of the reservoir fluid, and critical production rates. Also the effect of various
methods of vertical lift provided by production engineering studies should be

included in making the production forecast. A production schedule should yield a reasonable timing of production of the reserves over the life of the field.

These production rates are total oil, gas, or natural gas liquid rates, excluding water production. They are called gross production rates. The Company's working interest production rates must be determined since only the working interest provides the Company with income. The lease agreement should be examined to determine the amount of production reserved for royalty interest, overriding royalty, and so on.

The term *landowner's royalty* (usually abbreviated as royalty) is the interest of a party owning minerals in the ground where another party (the working interest) has gained the right to capture such minerals under a lease agreement. There are many specific forms of royalty. If only the usual 1/8 royalty applies, the working interest production would be 7/8 of the gross production.

4.3.2   Income from Oil and Gas Production

Revenue (to the Company) may be derived from future production estimates, by multiplying the working interest production by the appropriate crude oil or natural gas value. Future oil income usually employs the currently posted price of crude oil. Current prices are also employed for natural-gas liquids. However, in the gas of natural gas, existing contracts should be examined for the gas price to use.

4.3.3   Expenditures

The total expenditures incurred in producing oil and gas wells may be conveniently divided into two groups: (a) the total development costs and (b) the total operating costs.

Total Development Costs

This classification includes development costs such as drilling, well equipment, installation of pumping units, gathering lines, tank batteries, water- or gas-injection systems, pump stations, compressor stations, electrification of equipment, and recompletion jobs. It includes all expenditures required to drill and complete the well and to provide the well with production facilities required to lift the oil and gas from the bottom of the well to the surface and to transport it from the well head to the field terminal. Also included are all future expenditures required to keep the well on production (other than operating costs).

A summary of the expenditures included is as follows:

a.   Well costs (including casing).

b.   Well equipment (including well head connections, tubing, sucker rods, and subsurface pump).

c.   Depreciation (e.g., depreciation on drilling rigs and tools).

d. Overhead (excluding depreciation on overhead facilities).

e. Dry hole allowance (contingency costs).

f. Incumbent costs.

In order to calculate *contingency costs* (or dry hole allowance), a dry hole percentage must be estimated based on local experience, geological conditions, and reservoir type. The contingency costs are then calculated as the sum of items (a), (b), (c), and (d) multiplied by the factor

$$\frac{\text{dry hole percentage}}{100 - \text{dry hole percentage}}$$

The *incumbent costs* include the costs of production facilities and field construction (including overhead facilities) such as flowline, pumping unit, recompletions, per well share of block stations, gathering system, other field construction and facilities (including the overhead on these items), provided that these expenditures can be attributed to the drilling activity reviewed.

Total development costs should be classified as *tangible* or *intangible* expense to permit proper handling in the estimated income tax calculation. Intangible expenses are generally expenditures not represented by physical equipment. These items are not capitalized for tax purposes. Tangible expenses are costs of items of physical equipment and are tax deductible only by depreciation. Labor costs of installing equipment are a tangible expense except for those involved in drilling and completing wells and in installing subsurface pumps and sucker rods. In general, installation costs should be classified as tangible expense for surface equipment and intangible expense for subsurface equipment. The costs of the equipment is tangible expense. Table 4.2 is an example of a well development cost estimate.

Operating Costs

The expenditure incurred by producing an oil or gas well after it has been drilled and completed is called the operating cost. These costs may be further divided into two categories: (a) direct operating costs and (b) taxes. Total operating costs are the sum of direct operating costs and taxes, excluding income taxes.

a. *Direct operating costs* include all costs directly chargeable to individual leases except taxes. Labor and material costs for operation, maintenance, and repair are included. Also included are expenses for workovers, gas-processing plants, salt-water-disposal systems, and gas- or water-injection plants. These include an appropriate share of the overheads. Usually, future operating costs are estimated on a cost per well month or cost per lease month basis.

b. *Taxes* to which the Company is subjected on crude oil and natural gas production vary widely throughout the world. Severance, ad valorem, and

TABLE 4.2
Well Cost Estimate

Well: _____   **State:** _____

Prospect: _____   **County:** _____

Location: _____

Approximate Depth: ____3500'____

Intangible Costs	Dry Hole	Producer
Drilling: Move, rig up and out	$ 3,900	$ 3,900
Drill 5 days @ $3120/day	15,600	15,600
Test 1 day @ $3120/day	3,120	3,120
Completion 3 days @ $3120/day		9,360
Casing: Abandon 1 day @ $3120/day	3,120	
Surface 8⅝", 360' @ $5.85	2,250	2,250
Shoe, centralizers, etc.	200	200
Flange: Rental	100	—
Location:		
Survey	—	—
Rig site, cellar, conductor, and roads	2,000	2,000
Abandonment and clean up	500	500
Water:		
Pump: Haul	1,000	1,000
Lines	—	—
Storage	—	—
Fuel	—	—
Transportation:		
Casing: 8⅝", 5½"	150	750
Tubing: 2⅞"	—	150
Rods: ⅞", ¾"	—	—
Vacuum: Sump, completion fluid	1,000	1,000
Misc.:	500	500
Bits, reamers, core heads 12¼", 7⅞", 6¼"	2,600	2,800
Cementing services:		
Surface casing	1,500	1,500
Water string	—	2,500
Abandonment	2,000	—
Misc.	—	—
Mud:		
Drilling	3,500	4,000
Completion	—	500
Testing:		
Formation test	—	1,500
WSO test: Shoot and test	—	1,000
Production test	—	—
Coring	—	—
Logging:		
Electric log	1,500	1,500
Sidewall samples	1,500	1,500
Sonic—density	—	—

Intangible Costs	Dry Hole	Producer
Dipmeter	—	—
Cement bond—gamma log	—	—
Mud log	1,500	1,500
Misc.:	—	—
Perforating:		
WSO	—	—
Shop: Liner	—	400
Gun	—	—
Services and supervision:		
Geological	1,000	1,500
Engineering	2,400	4,000
Crop damage	—	—
Welding	300	600
Misc.: Casing, tongs, and tools	325	775
Monel 5 days @ $21/day	105	105
Survey instrument	450	450
Tubing tools, power tongs		600
Kelly collars, etc.		350
Total intangible costs	$ 52,120	67,410
Contingencies	$ 5,300	6,800
Total estimated costs	$ 57,420	74,210

Tangible Costs		
Casing:		
Intermediate	$ —	—
Water: 3500', 5½", 17#	—	19,000
Misc.:	—	—
Casing equipment:		
Intermediate	—	—
Water: Solid baffle collar, filtrol collar,		
centralizers, etc.	—	1,500
Tubing: 2⅜"	—	6,700
Rods: ¾"	—	3,150
Pump	—	450
Well head connections	—	1,500
Valves and fittings	—	1,500
Pumping unit and engine	—	6,500
Flow lines	—	1,000
Tank farm	—	20,000
Labor	—	2,000
Misc.: Move tanks	—	2,000
Total tangible costs	—	65,300
Contingencies	—	6,500
Total costs	—	71,800
Total intangible and tangible costs	$ 57,420	$146,010

windfall profit taxes are common in the United States, as are federal and state income taxes. In other parts of the world the basis and rates of taxation vary considerably and must be analyzed separately in each case.

Severance taxes include state production taxes and conservation taxes. They are levied by state governments and are based on production volume, production value, or both. The Company pays taxes only on its working interest production. The tax rates differ from state to state.

Ad valorem taxes are property taxes levied by state, county, and municipal governments. The taxes may be based on the reserves in the ground, actual average daily production rate, or other factors. The Company pays taxes only on its working interest in the property. The tax rates and method of calculation vary widely from one area to another.

Windfall Profit Tax

The Windfall Profit Tax Act of 1980 became effective March 1, 1980. The tax is applied to each barrel of oil removed and sold from the premises, with oil used on the premises from which it is produced not being subject to tax. The Windfall Profit Tax is an excise tax and, therefore, a deductible expense in computing income tax. Full percentage depletion is allowed on the windfall profit for both independent producers and royalty owners.

Generally there are three tax tiers that include all crude oil production except: (a) certain newly discovered Alaskan production, (b) front end tertiary oil, and (c) production owned by state and local governments, Indian tribes, charitable medical institutions, and educational institutions. The windfall profit tax depends on the tier level of crude oil. The three tax tiers are:

Tier I Includes all crude oil which for the U.S. Department of Energy pricing purposes is classified as lower tier, upper tier, market level new crude oil, marginal property production, and heavy oil (between 16° and 20° API) that does not meet the heavy oil definition for tax purposes. Integrated companies and all royalty owners are taxed at 70% while independent producers are taxed at 50%.

Tier II Includes stripper well crude oil and national petroleum reserves crude oil. Integrated companies and all royalty owners are taxed at 60% while independent producers are taxed at 30%.

Tier III Includes newly discovered crude oil, heavy crude oil (16°API or less), and incremental tertiary crude oil. All producers are taxed at 30%.

The base price for a barrel of crude oil produced in the United States depends upon whether the oil is characterized as Tier I, II, or III. The base price for each tier is subsequently adjusted quarterly for the effect of post-June 30, 1979 inflation.

The windfall profit tax rate will be phased out at 3 percent a month over a 33-month period beginning the later of January 1, 1988 or the month after cumulative net revenues raised by the tax have exceeded $227.3 billion. In all events the phase out is scheduled to begin by January 1991.

In all cases, the method of computing the tax is essentially the same. The elements necessary for calculating the tax are:

1. The removal price which is the selling price.

2. Adjusted base price.

3. Severance tax adjustment.

4. Windfall profit tax rate.

5. The number of barrels sold.

There are two steps in the computation to the windfall profit tax. The first step is determination of the windfall profit:

Windfall profit (WP) = *Selling Price* (SP)
    − *Adjusted Base Price* (ABP)
    − *Severance Tax Adjustment* (STA)

The selling price or removal price is the price at which a producer sells crude oil to the first purchaser at the wellhead. Presently the selling price of crude oil in the United States has been decontrolled.

The second step is computation of the windfall profit tax:

Windfall Profit Tax (WPT) = *the lower of windfall profit or the net income limitation times the windfall profit tax rate* (WTR)

The net income limitation states that the amount of windfall profit subject to tax cannot exceed 90 percent of the net income per barrel from each separate property for the taxable year.

In equation form, the windfall profit tax is given by:

$$WPT = [(SP - ABP) - STA] \times WTR$$

or

$$WPT = (SP - ABP) \times (1 - STR) \times WTR$$

where *STR* = severance tax rate.

Example 4.1.
Windfall Profit Tax Computation

Selling Price *(SP)*	$ 30.00
Adjusted Base Price *(ABP)*	(16.00)
	$ 14.00

Severance Tax Adjustment *(STA)*	
(14 × 0.125)	(1.75)
Windfall Profit *(WP)*	12.25
Windfall Profit Tax Rate *(WTR)*	× 0.30
Windfall profit tax *(WPT)*	$ 3.68

The windfall profit tax is $3.68 per barrel of crude oil removed and sold. For a state with no deductible severance tax, the calculation would be:

$$WPT = \$(30 - 16) \times 0.30 = \$4.20 \text{ per barrel}$$

Computation of net income for purposes of windfall profit tax is in accordance with special rules and definition. The net income limitation is applied on a *yearly* basis rather than a quarterly basis. The net income attributable to a barrel is determined by dividing taxable income from the property, which is attributable to taxable oil, by the number of barrels of that oil produced from the property during the taxable year. The computation of the net income limitation proceeds as follows:

    Oil Revenue
    − Oil severance tax
    − ad valorem tax
    − pro rata share of operating cost
    − tertiary injectants (optional)
    − pro rata share of depreciation expenses
    − imputed cost depletion
    = *WPT* net income
    × 0.90
    = 90% *WPT* net income
    ÷ net oil production
    = 90% net income limitation per barrel
    × *WPT* rate
    = *WPT* 90% limit, $/bbl
    × barrels of windfall oil
    = *WPT* 90% limit due

Overheads

Overheads are expenditures incurred on behalf of the operations in general. There is no straightforward relationship between overheads and operating costs and development costs. Overheads are allocated to the various phases of activity by means of simple rules that vary from company to company. There may be cases where simple rules, like expressing overheads by means of a percentage on top of direct costs, will not be justified. Capital expenditure may sometimes qualify for inclusion under overheads (e.g., the cost of building a central office). It is recommended that full details be given as to the methods in which overheads have been accounted for.

Items of overhead cost include the following:

Plant overhead

Home office overhead

Administration and management

Research and development

Selling expense

Debt interest

Royalties

Property insurance and local taxes

Bonuses, awards, other incentives

Total Expenditures Before Income Tax

Total expenditures before income tax are the sum of total development costs and total operating costs.

4.3.4   Income Taxes

The net revenue before income tax (or net revenue after expenses) is determined by subtracting total expenditures before income tax from total revenue. The state and federal income taxes and capital investment are subtracted from the net revenue before income tax to yield after tax net cash flow.

State Income Tax

The state income tax is obtained by multiplying the net revenue before income tax by the appropriate state income tax rate. These tax rates differ considerably from one state to another. The state income taxes are low compared to the federal income tax.

Federal Income Tax

The federal income tax is obtained by multiplying the taxable income after all deductions by the corporate tax rate, which may be as high as 46 percent. The corporate tax rate is progressive, varying with the taxable income. For illustrative purposes a corporate tax rate of 50 percent is assumed, unless otherwise stated. The state income tax is a deductible expense for federal income tax calculations.

The basic method for calculating United States federal income tax is as follows:

1. Total Revenue

   minus

2. Deductions

   Total operating costs
   State income tax
   Intangible development costs
   Depreciation

   equals

3. Net Income Before Depletion Allowance

   minus

4. Depletion Allowance
   At 50% of Net Income Before Depletion Allowance, or
   At a percentage of Total Revenue, or
   At cost

   equals

5. Net Taxable Income

   and

6. (Tax Rate) × (Net Taxable Income)

   equals

7. Federal Income Tax

Example 4.2. Cash Flow Generation. Consider an oil field to be developed. For a first year output of 3.6 million barrels of crude oil sold at a net price of $24.00 per barrel, the gross revenue is $86,400,000. One-eighth of this amount is paid as royalty to the landowner, leaving a net gross income—the working interest—of $75,600,000 for the year under consideration. This revenue and cost items are listed in Table 4.3 for the first two years of production.

TABLE 4.3
Cash Flow Generation (Values in Thousands)

	Year 1	Year 2
	Barrels	
Gross production	3,600	1,800
Working interest production	3,150	1,575
	Dollars	
Working interest revenue	75,600	37,800
Operating costs	(18,000)	(9,000)
Overhead	(8,000)	(8,000)
State taxes	(7,600)	(3,800)
Depreciation	(7,500)	(7,500)
Expensed investment	(12,000)	0
Depletion	(8,000)	(4,000)
Taxable income	14,500	5,500
Federal income tax at 50%	(7,250)	(2,750)
Net profit	7,250	2,750
Depreciation	7,500	7,500
Depletion	8,000	4,000
Expensed investment	12,000	0
Investment tax credit	1,000	0
Capital investment	(22,000)	0
After tax net cash flow	(1,250)	14,250

4.4 CUMULATIVE CASH FLOW DIAGRAM

The flow of funds into and out of a project during its life may be shown in a cumulative cash flow diagram, Fig. 4.2. At the start of a project its cash position is zero, A. For economic evaluation the start of a project is defined as the time when the first cash flow occurs. Expenditure is a negative cash outflow and income a positive cash inflow.

At the beginning, after the decision is made that the project has economic merit, the firm's bank secures the necessary capital and then makes it available to the project as needed for the purchase of land, design, and other preliminary work, which all cost money (A-B). This is followed by the main phase of capital investment in well drilling, buildings, plant and equipment (B-C). Working capital is then spent and the project is commissioned in readiness for commercial production, which gets under way at D. The working capital is needed to support start-up and to fund the going operation until sales dollars come in.

The lowest point D on the curve represents the maximum cumulative debt incurred by the project. Beyond this point in time, income from sales of the product being made exceeds the operating and other costs and so the cumulative cash flow curve turns upward. The cumulative cash flow is plotted throughout the

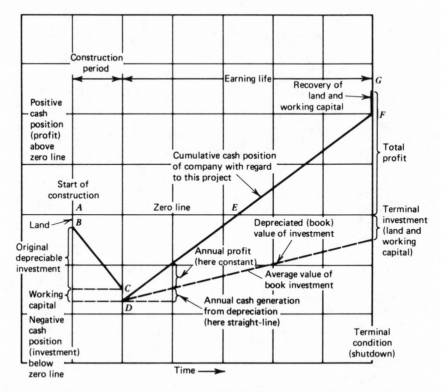

Figure 4.2 Cumulative cash position chart. This is also termed a "cash flow chart" (after Uhl and Hawkins).

life of the project and, at the end of the project life, adjustments are made for the recovery of land and working capital and also for salvage value, if any.

The cumulative cash flow diagram proves to be useful in following the financial history of a project. It also provides a powerful device to readily understand the common *profitability criteria*, which is the subject of Chapter 6. Because at the time an economic evaluation is made all cash flows are in the future, they are subject to varying degrees of uncertainty. Initially this uncertainty and its effects are ignored. Its important implications for economic evaluation and decision making are considered in later chapters.

4.5 DEPRECIATION, DEPLETION, AMORTIZATION, AND TAX CREDITS

Depreciation, depletion, and amortization are the means of recovering your investment in certain types of property in before tax dollars. In addition to these many governments encourage investment in capital equipment by giving an investment tax credit.

4.5.1 Depreciation

Depreciation is the process of prorating as expenses the capitalized costs of certain items over a period of years. Depreciation is a noncash charge, deductible from the tax base, which represents a reasonable allowance for the exhaustion, wear and tear, and obsolescence of depreciable property used in business or held for the production of income. This enables a firm or business to recover the cost of a depreciable asset during its estimated useful life. Depreciable items normally include well equipment (tangible drilling and development costs), and surface equipment. In general, an item can be declared depreciable if it retains a usable or reusable value over a period of time.

The above definition indicates that a given depreciable item will have some value if it is not utilized to the exhaustion of its useful life. The *salvage value* of an item then is the estimate at the time the asset is acquired of the amount of money that will be realized upon its sale or other disposition for some other use.

There are several methods of computing regular depreciation allowance. The methods most generally used are: (1) the straight-line method; (2) the declining balance method; (3) the sum of the years digits method; (4) a combination of declining balance and straight-line methods; and (5) the units of production method. Any reasonable method of depreciation is acceptable by the federal government for income tax computations, provided that it is consistently applied by the firm. In addition, some assets qualify for special deductions such as accelerated first-year depreciation.

Straight Line

The straight-line method of depreciation allows the deduction of an equal amount from the taxable income each year over the life of the asset. The procedure is to estimate the useful life of a piece of equipment and its salvage value, if any, at the end of such useful life. The formula for the straight-line depreciation allowance is

$$D = \frac{C - S}{n} \tag{4.1}$$

where

$$D = \text{depreciation allowance per year}$$

$$S = \text{estimated salvage value}$$

$$n = \text{estimated asset life in years.}$$

The straight-line depreciation rate is $1/n$.

Declining Balance

The declining balance depreciation methods allow calculation of the annual depreciation at 150% or 200% of the straight-line rate applied to the *undepre-*

ciated balance of the asset, which is the book value. Since the salvage value is not deducted from the cost of the depreciable equipment, however, the total depreciation cannot exceed the total cost of the asset less its salvage value. The formula for the declining balance method is

$$D = \frac{F}{n} (C_u) \tag{4.2}$$

where

$$F = \text{depreciation factor}$$

$$C_u = \text{undepreciated balance for a given year}$$

If $F = 2$, the depreciation rate is $2/n$ (double the straight-line rate) and we obtain the *double declining balance* depreciation method. The double declining balance rate is the highest declining balance rate permissible and, thus, is the most widely used in the industry.

Although the rate is constant, it is applied to a declining balance each year. A larger depreciation deduction is taken for the first year and a gradually reduced deduction is taken in subsequent years.

Sum of the Years Digits

The sum of the years digits uses a declining rate per year that is applied to the total costs of the asset less the salvage value. The rate is determined by a fraction, with the numerator of the fraction being the remaining life of the equipment as of the beginning of the year, while the denominator is the sum of the digits in the total life of the asset. The formula for the sum of the years digits method is

$$D = \frac{(n - m + 1)}{\sum\limits_{i=1}^{n} m} (C - S) \tag{4.3}$$

where m is the life of asset up to the time of calculation in years.

Table 4.4 summarizes the depreciation calculations for the three methods just discussed and gives formulas for the book value at the end of the year.

Units of Production

The unit of production depreciation method is based on equipment usage. A prerequisite is a reasonably accurate estimation of the recoverable reserves. The depreciation is charged at a fixed amount per unit produced during the accounting period. In the case of crude oil production, the total tangible expenditures (less salvage value) are divided by the total estimated net working interest barrels of oil to be recovered, thus determining the depreciation per barrel. This value can then

TABLE 4.4
Formulas for Depreciation and Book Value

Method	Depreciation Rate	Operative Upon	Annual Amount Depreciated	Book Value at End of Year	Is Salvage Value Taken Into Account?
Straight line	$\dfrac{1}{n}$	$(C - S)$	$\dfrac{1}{n}(C - S)$	$C - \left(\dfrac{C - S}{n}\right)m$	Yes
Declining balance	$\dfrac{2^a}{n}$	$C\left(1 - \dfrac{2}{n}\right)^{m-1\,b}$	$\dfrac{2}{n}C\left(1 - \dfrac{2}{n}\right)^{m-1}$	$C\left(1 - \dfrac{2}{n}\right)^m$	No
Sum of the digits	$\dfrac{n - m + 1}{\displaystyle\sum_1^n m}$	$(C - S)$	$\dfrac{(n - m + 1)(C - S)}{\displaystyle\sum_1^n m}$	$C - (C - S)\left[\dfrac{\displaystyle\sum_1^n (n - m + 1)}{\displaystyle\sum_1^n m}\right]$	Yes

C is the original investment, which includes installation costs.
S is the estimated salvage, which is net from estimated sale value minus dismantling cost.
n is the estimated asset life in years.
m is life of asset up to the time of calculation in years.
[a] The "practical" maximum rate allowable by the Internal Revnue Service. Practically no one uses a lesser value.
[b] This is the book value at the end of the previous year.

be applied to the estimated production for each time period to determine the depreciation for the period. The formula for the units of product depreciation method is

$$D = \frac{(C - S)P}{R}$$

(4.4)

where

$$R = \text{total recoverable reserves}$$

$$P = \text{production for a given year}$$

This method is used frequently for depreciating oil, gas, and mineral property depreciable assets.

Example 4.3. A piece of oil production equipment with an estimated life of 5 years is purchased for $33,000. Its salvage value at the end of the fifth year is estimated to be $3000. It has the following production profile:

Year	Production (thousands of barrels)
1	500
2	2,000
3	1,300
4	800
5	400
	5,000

In Tables 4.5 to 4.8 the annual depreciation write-offs are compared using the four methods discussed. Although the cumulative depreciation over the lifetime of the asset is the same in all four methods, the distribution of the allowances in time varies and has different impacts on annual cash flows and on the profitability of the project.

In Table 4.6 only $1277 allowance is permissible during the fifth year since the cumulative depreciation allowance cannot exceed the equipment cost less its salvage value.

Combination of Declining Balance and Straight-Line

For tax purposes, it is advantageous to use the accelerated depreciation methods whenever possible. This benefits the project by increasing depreciation in the early years. An important rule to follow is to take the biggest deduction possible at the earliest date possible.

It is permissible to switch from one depreciation method to another anytime during the useful life of an asset. The switch should be made in the year when the depreciation allowance by the method used becomes less than it would be by another method. One common switch is from double declining balance depreciation to straight-line depreciation. The procedure is illustrated by Example 4.4.

TABLE 4.5
Straight-Line Method

Year	Cost Less Salvage ($)	Annual Rate (%)	Annual Allowance ($)	Cumulative Allowance ($)
1	30,000	1/5 = 20	6,000	6,000
2	30,000	20	6,000	12,000
3	30,000	20	6,000	18,000
4	30,000	20	6,000	24,000
5	30,000	20	6,000	30,000

TABLE 4.6
Double Declining Balance Method

Year	Unrecovered Cost ($)	Rate (%)	Annual Allowance ($)	Cumulative Allowance ($)
1	33,000	2/5 = 40	13,200	13,200
2	19,800	40	7,920	21,120
3	11,880	40	4,752	25,872
4	7,128	40	2,851	28,723
5	4,277	40	1,277	30,000

TABLE 4.7
Sum of the Years Digits Method

Year	Cost Less Salvage ($)	Rate	Annual Allowance ($)	Cumulative Allowance ($)
1	30,000	5/15	10,000	10,000
2	30,000	4/15	8,000	18,000
3	30,000	3/15	6,000	24,000
4	30,000	2/15	4,000	28,000
5	30,000	1/15	2,000	30,000

TABLE 4.8
Units of Production Methods

Year	Cost Less Salvage ($)	Rate	Annual Allowance ($)	Cumulative Allowance ($)
1	30,000	5/50	3,000	3,000
2	30,000	20/50	12,000	15,000
3	30,000	13/50	7,800	22,800
4	30,000	8/50	4,800	27,600
5	30,000	4/50	2,400	30,000

Example 4.4. Using the information from Example 4.3, calculate the double declining balance depreciation allowance each year, but swtich to the straight-line depreciation when advantageous.

Year	Method	Rate	Unrecovered Cost ($)	Depreciation ($)
1	DDB	2/5	33,000	13,200
2	DDB	2/5	19,800	7,920
3	DDB	2/5	11,880	4,752
4	SL	1/2	7,128	3,564
5	SL		564	564

A switch to a straight-line depreciation is advantageous in year 4 because it gives a larger deduction than would be allowable if we did not switch. Only a $564 allowance is permissible during the fifth year since the cumulative depreciation allowance cannot exceed $30,000.

The general depreciation method used in the petroleum industry is an 11-year life with a sum of the years digits for the first two years, followed by a double declining balance.

4.5.2   Depletion

Depletion is the exhaustion of a natural resource as a result of its extraction. As oil and gas or some other natural resources are used up, a depletion reserve is set up in recognition of the fact that, as the company sells its product it must get back not only the cost of extracting it but also the original cost of attaining it. Hydrocarbons and mineral resources are termed *nonrenewable* because they are created by a geologic process over long geologic periods of time.

A depletion allowance represents the amount allowed as an annual deduction in arriving at the net income for the taxable year from mineral properties. Only an operating owner or an owner of an economic interest may claim depletion deduction. Thus, royalty owners may claim the allowance; shareholders may not. The two permissible methods of computing depletion currently are (1) cost depletion and (2) percentage depletion. The taxpayer must deduct the larger of cost depletion and percentage depletion. The amount that may be deducted from taxable income is known as *allowable depletion*.

Cost Depletion

Cost depletion basis for minerals includes the cost of acquiring a mineral property and the exploration expenditures incurred in discovering this deposit. Mineral rights acquisition costs, lease bonuses, and other equivalent ascertained costs including geological and geophysical survey costs, legal costs, and assessment costs are the primary costs that go into the cost depletion cost basis.

Cost depletion is computed by dividing the total number of recoverable units in the deposit at the beginning of a year (determined in accordance with prevailing industry methods) into the *adjusted basis* of the property for that year, and multiplying the resulting rate per unit by (1) the number of units for which payment is received during the tax year, if the cash receipts and disbursement method is used, or (2) the number of units sold if an accrual method of accounting is used. The adjusted basis of the property is the original cost of the property less all the depletion (cost and percentage) allowed on the property since its acquisition by the taxpayer. The formula for calculating cost depletion is similar to that for units of production depreciation:

$$CD = B \left(\frac{P}{R + P} \right) \tag{4.5}$$

where

CD = annual cost depletion allowance

B = the "adjusted basis" of the property

P = units of production sold or for which payment was received during the tax year

R = "recoverable units" of production remaining at the end of the tax year.

Example 4.5.   The purchase price of a producing ore property was $3,000,000 and exploration and development costs included in the cost depletion basis amounted to $2,000,000. Engineering estimate of the recoverable reserves is 4,000,000 barrels. The yearly depletion for three years in the life of the project is as follows:

Year	Production (Barrels)	Adjusted Basis ($)	Annual Cost Depletion ($)
1	400,000	5,000,000	500,000
2	200,000	4,500,000	250,000
3	150,000	4,250,000	187,500

Adjusted basis for first year = $3,000,000 + $2,000,000 = $5,000,000

$$\text{Cost depletion per barrel} = \frac{\$5,000,000}{4,000,000 \text{ bbl}} = \$1.25/\text{bbl}$$

Similarly, in the second year, the cost depletion per barrel is

$$= \frac{\$5,000,000 - \$500,000}{(4,000,000 - 400,000) \text{ bbl}} = \$1.25/\text{bbl}$$

TABLE 4.9
Allowable Percentage Depletion Schedule

Year	Average Daily Production of Oil (bbl/day)	or	Average Daily Production of Gas (cf/day)	Rate (%)
1975	2,000		12,000,000	22
1976	1,800		10,800,000	22
1977	1,600		9,600,000	22
1978	1,400		8,400,000	22
1979	1,200		7,200,000	22
1980	1,000		6,000,000	22
1981	1,000		6,000,000	20
1982	1,000		6,000,000	18
1983	1,000		6,000,000	16
1984 and years thereafter	1,000		6,000,000	15

Percentage Depletion

Percentage depletion permits the deduction of a percentage specified by law, depending on the mineral involved, of gross income from a mineral property. The deduction for percentage depletion must not exceed 50 percent of taxable income from the property after all deductions except depletion have been taken. Unlike depreciation and cost depletion it needs no starting basis even if all costs have been recovered. Percentage depletion allowance plays an important role in the development of mineral resources by providing the operating owners with a relatively inexpensive internal sources of finance for further investment in exploring and developing new reserves.

The U.S. Congress has approved a graduated phaseout of percentage depletion allowance for large domestic oil companies. Percentage depletion is now allowed only to relatively small independent producers and royalty owners as indicated in Table 4.9. For the small producer the Congress allowed a 22 percent rate through 1980 with 2 percent annual rate decreases until 1984. A rate of 15 percent is allowed in 1984 and years thereafter. Production rate limits started at 2000 barrels per day of oil or 12 million cubic feet per day of gas and declined 200 barrels per day of oil or 1.2 million cubic feet per day of gas until 1980. Production allowables have been fixed at 1000 barrels per day of oil or 6 million cubic feet per day of gas after 1980. Where the taxpayer has production of both oil and natural gas, the allowable production must be divided between the two types of production. One barrel of crude oil is equivalent to 6000 cubic feet of natural gas.

Example 4.6. Here is a demonstration of the computation of a depletion allowance assuming a percentage depletion of 15 percent.

	Property		
	A	**B**	**C**
Gross production, bbl	150,000	160,000	120,000
WI production, bbl	131,250	140,000	105,000
Gross income	$3,937,500	$4,200,000	$3,150,000
Operating costs	(1,200,000)	(2,400,000)	(2,000,000)
Depreciation	(600,000)	(650,000)	(550,000)
Taxable income before depletion	$2,137,500	$1,150,000	$ 600,000
Depletion allowances			
Cost depletion	435,000	450,000	425,000
Percentage depletion at 15%	590,625	630,000	472,500
50% of taxable income	1,068,750	575,000	300,000
Allowable depletion	(590,625)	(575,000)	(425,000)
Taxable income	$1,546,875	$ 575,000	$ 175,000

4.5.3 Amortization

Amortization applies to certain capital expenditures. Some of the capital expenditures that may quality for amortization for federal income tax purposes are: (1) organization expenses of a company, (2) research and development expenses, (3) trademark and trade name expenditures, (4) premiums paid on partially and fully tax-exempt bonds, and (5) cost of acquiring a lease for business purposes (other than a mineral lease). The firm may deduct each year a proportionate part of these expenditures and thus recover them in a manner similar to straight-line depreciation.

4.5.4 Investment Tax Credit

A credit against income tax is allowed for qualified investment in certain property. Such property should be depreciable, have a useful life of three or more years, and be placed in service during the year. The investment tax credit amounts to 10 percent of the qualifying asset cost subject to the credit. The credit may be taken on no more than $100,000 of the costs of qualifying *used* property in any tax year.

The investment tax credit is not a deduction from taxable income. It is a one-time credit against your tax bill.

4.5.5 Capital Gains Tax

When a capital asset is sold, you may have a capital gain or loss. A *capital gain* or *capital loss* is the difference between the amount realized from a sale or exchange of a capital asset and the adjusted basis of the asset transferred.

For companies, capital gains may be taxed either as ordinary income at the regular tax rate or at the alternate rate (such as 28%). A company may select the method that yields the smallest tax. If a depreciable asset is sold for more than its book value, tax must be paid on the gain.

4.5.6 Summary

The computation of after federal income tax net cash flow can be summarized as follows:

$$
\begin{aligned}
\text{After federal income tax, net cash flow} = \ & (1 - T)\,(\text{revenue}) \\
& - (1 - T)\,(\text{state taxes}) \\
& - (1 - T)\,(\text{operating costs}) \\
& - (1 - T)\,(\text{overhead}) \\
& - \text{capitalized investment} \\
& + (T)\,(\text{depreciation}) \\
& - (1 - T)\,(\text{expensed investment}) \\
& - \text{bonus and leasehold costs} \\
& + (T)\,(\text{depletion}) \\
& + (C)\,(\text{capitalized investment}) \\
& + \text{sales price} \\
& - (G)\,(\text{sales price} - \text{book value})
\end{aligned}
$$

where

$$T = \text{corporate income tax rate}$$

$$C = \text{investment tax credit rate}$$

$$G = \text{capital gains tax rate}$$

REFERENCES

Hughes, R. V.: *Oil Property Valuation*, Robert E. Krieger, New York, 1978.

Roebuck, I. F., Jr.: *Economic Analysis of Petroleum Ventures*, Institute for Energy Development, Oklahoma City, 1979.

Rudawsky, O.: "Economic Feasibility Studies in Mineral and Energy Industries," Colorado School of Mines Mineral Industries Bulletin, Vol. 20, Nos. 3 and 4, 1977.

Stermole, F. J.: *Economic Evaluation and Investment Decision Methods*, Third Edition, Investment Evaluations Corporation, Golden, Colorado, 1980.

Uhl, V. W. and Hawkins, A. W.: *Technical Economics for Engineers*, American Institute of Chemical Engineers Continuing Education Series 5, American Institute of Chemical Engineers, New York, 1971.

5

TIME VALUE OF MONEY

5.1 INTRODUCTION

A basic concept in economic analysis is that money has a time value. This can be stated as: a sum of money *now* is normally worth more than an equal sum of money at some *future* date. The long lead time between initial investment of funds in exploration and development of oil and gas resources and the inflow of revenue when these fields are fully operative necessarily implies that we are dealing with different values of money because of the different time aspect.

The techniques in common use for evaluating projects and comparing alternatives require recognition of the time value of money and understanding of the methodology for handling this concept, which is the proper use of interest formulas and tables.

Interest is the difference between the value (price) of an earlier rather than later availability of funds. It is the price we pay for the use of money on loan. The amount of the loan on which interest is paid is called the *principal*. The *interest rate* is the fraction of the principal that is paid per unit of time. The interest rate is usually denoted by the letter, i. It is almost invariably positive because present money is at a premium compared with future money. For convenience, i, is usually expressed as a percent per unit time (e.g., 10% per year).

The actual interest rate depends on a number of circumstances among which the most important is the cost of capital that depends on (1) the risk involved, (2) availability of capital, (3) size of loan, (4) duration of loan, and (5) whether the return is taxable or not. Other factors particularly important for commercial ventures are the nature of the industry and the position of the firm in the business world.

Inflation should not be confused with interest. Inflation is due to a general rise in the price level (you can buy less with the same number of dollars). However, the inflationary effect augments and magnifies the time value of money. Changes in price levels do not create the time value of money; they only influence its magnitude for any given time period.

5.2   INTEREST FORMULAS

Conventional interest tables are based on the premise that interest is compounded periodically and that income, or payment, is received at the end of each period. Common ways that compound interest may be expressed are shown in the following sections. The symbols used for interest formulas are:

i = interest rate per interest period

n = number of interest periods

P = present sum of money

F = future sum of money n interest periods from present date, equivalent to P with interest rate i

A = end of period payment in a uniform series continuing for the coming n periods, the entire series equivalent to P at interest rate i

Single Payment Compound Amount Factor

The amount of compound interest is often called the *single payment compound amount*. It shows how $1 at compound interest will grow. If a capital sum P (the principal) is placed on compound interest at a rate i compounded annually,

$$\text{Interest for first year} = iP$$
$$\text{Sum at end of first year} = P + iP = P(1 + i)$$
$$\text{Interest for second year} = iP(1 + i)$$
$$\text{Sum at end of second year} = P(1 + i) + iP(1 + i) = P(1 + i)^2$$
$$\text{Similarly, for } n \text{ years}$$
$$F = P(1 + i)^n \tag{5.1}$$

The factor $(1 + i)^n$ is the single payment compound amount factor.

Example 5.1.   If a sum of $25,000 is invested at the interest rate of 8 percent per year for 10 years, what will it grow to?

Solution
This is a single payment compound amount problem.

$$P = \$25,000$$
$$i = 8\%/\text{year}$$
$$n = 10 \text{ years}$$
$$F = ?$$

The single payment compound amount factor is:

$$(1 + 0.08)^{10} = 2.1589$$

The compound amount then becomes

$$F = \$25,000(2.1589) = \$53,972.50$$

It should be noted that the compound amount factor will vary with the time and frequency of compounding. In Example 5.1, the compounding frequency is once a year at year-end. Other possible compounding frequencies may be once a year at midyear, semiannually, quarterly, or continuously.

Single Payment Present Value (Present Worth) Factor

The *present value* or *present worth* is the worth today of $1 due sometime in the future. It is the amount that must be deposited at compound interest today to grow to $1 in the future. If $1 deposited today will grow to F after n periods, then $(1/F)$ deposited at compound interest today will grow to $1 in the future. Thus present value or present worth is the reciprocal of compound amount. Expressed as an equation:

$$P = F\left[\frac{1}{(1 + i)^n}\right] \tag{5.2}$$

The factor $1/(1 + i)^n$ is the single payment present value factor.

Example 5.2. To receive $10,000 in the future 20 years from now, how much should I deposit today at 10 percent interest per year?

Solution
This is a present value problem.

$$P = ?$$
$$F = \$10,000$$
$$i = 10\%$$
$$n = 10 \text{ years}$$

The single payment present value factor is

$$1/(1.10)^{20} = 0.1486$$

The present value is

$$P = \$10,000(0.1486) = \$1,486$$

This process of converting future income into present value or present worth is called *discounting*. The single payment present value factor is also called the *discount factor*. Since immediate availability of resources is usually preferred over future availability, the present value of future income is somewhat smaller.

The previous comment about the frequency and timing of compounding during a discrete time period applies also in the case of discounting. The only difference is that the higher the frequency (and the shorter the time period), the lower the present value.

Sinking Fund Factor

A sinking fund schedule of payments shows the amount that must be deposited periodically to grow to $1 in the future. If the amount A received at the end of the first year is reinvested immediately, it earns interest for $(n - 1)$ years at i, and its future amount is $A(1 + i)^{n-1}$. The second payment A reinvested amounts to $A(1 + i)^{n-2}$ at the end of n years, etc. The last payment A at the end of the nth year earns no interest. Therefore, the sum

$$F + A[1 + (1 + i) + (1 + i)^2 + \cdots + (1 + i)^{n-1}] \qquad (5.3)$$

This can be simplified by multiplying each side by $(1 + i)$,

$$F(1 + i) = A[(1 + i) + (1 + i)^2 + (1 + i)^3 + \cdots + (1 + i)^n] \qquad (5.4)$$

and subtracting Eq. (5.3) from Eq. (5.4):

$$Fi = A[(1 + i)^n - 1]$$

or

$$A = F \left[\frac{i}{(1 + i)^n - 1} \right] \qquad (5.5)$$

Example 5.3.   Calculate the uniform series of equal payments made at the end of each year for 12 years that are equivalent to a $10,000 payment 12 years from now if interest is 9 percent per year compounded annually.

Solution

$$F = \$10,000$$

$$i = 9\% \text{ per year}$$

$$n = 12 \text{ years}$$

$$A = ?$$

Sinking fund factor is

$$\frac{0.09}{(1.09)^{12} - 1} = 0.04965$$

$$A = \$10,000(0.04965) = \$496.50$$

Capital Recovery Factor

This factor is equivalent to the amount that must be received periodically for n periods to be worth $1 today. It shows the periodic payment necessary to pay off a loan of $1, or the amount of periodic payments that could be obtained for n periods from a $1 investment.

The formula may be derived from Eq. 5.5. Substituting $S = P(1 + i)^n$, we get

$$A = P(1 + i)^n \left[\frac{i}{(1 + i)^n - 1} \right] = P \left[\frac{i(1 + i)^n}{(1 + i)^n - 1} \right] \tag{5.6}$$

or

$$A = P \left[\frac{i}{(1 + i)^n - 1} + i \right] \tag{5.7}$$

The difference between the sinking fund and capital recovery should be noted. The sinking fund represents the amount that must be set aside periodically to grow to $1 in the future. The capital recovery represents the amount that must be set aside periodically to grow to a future amount whose present value is $1.

Example 5.4. To retire a present loan of $10,000 with interest at 9 percent per year over 12 years, what should the annual payments be at the end of each year?

Solution

Capital recovery factor

$$= \left[\frac{0.09}{(1.09)^{12} - 1} + 0.09 \right] = 0.13965$$

$$P = \$10,000$$

$$i = 9\% \text{ per year}$$

$$n = 12 \text{ years}$$

$$A = \$10,000(0.13965) = \$1396.50$$

Uniform Series Compound Amount Factor

This factor shows how $1 deposited periodically at compound interest will grow. All deposits are made at the end of a period. It is the reciprocal of the sinking fund factor. Expressed as an equation:

$$F = A \left[\frac{(1 + i)^n - 1}{i} \right] \tag{5.8}$$

Example 5.5. Calculate the future compound amount 10 years from now of a uniform series of $10,000 incomes received at the end of each year for the next 10 years if interest is at 11 percent compounded annually.

Solution

$$A = \$10,000$$

$$i = 11\% \text{ per year}$$

$$n = 10 \text{ years}$$

$$F = ?$$

Uniform series compounded amount factor

$$= \left[\frac{(1.11)^{10} - 1}{0.11} \right] = 16.722$$

$$F = \$10,000(16.722) = \$167,220$$

Uniform Series Present Value (Present Worth) Factor

This factor shows what $1 received periodically for n future periods is worth today. It is the reciprocal of the capital recovery factor. Expressed as an equation,

$$P = A \left[\frac{(1 + i)^n - 1}{i(1 + i)^n} \right] \tag{5.9}$$

The uniform series present value factor is useful in evaluating investments that produce a constant yearly income. The factor is numerically equal to the time required to return an investment (the payment time). Thus, for a constant yearly income received for n years,

$$\text{Payout time} = \frac{(1 + i)^n - 1}{i(1 + i)^n} \tag{5.10}$$

Example 5.6. An example of the use of the annuity formula is a project that is estimated to generate a constant annual cash flow of $20,000 for 10 years. If the rate of interest is 12 percent compounded annual, what should be the maximum level of investment at the beginning of the project?

Solution

The present value of these annuities is

$$P = \$20,000 \left[\frac{(1.12)^{10} - 1}{0.12(1.12)^{10}} \right]$$

$$= \$20,000(5.650) = \$113,000$$

When the series of annuities is expected to go on perpetually, Eq. 5.9 becomes

$$P = \lim_{n \to \infty} A \left[\frac{1 - (1 + i)^{-n}}{i} \right]$$

or

$$P = \frac{A}{i} \tag{5.11}$$

As n tends toward infinity $(1 + i)^n$ tends to zero. This case is referred to as a *perpetuity*.

Using data from Example 5.6 for calculating the present value of future constant income for a perpetuity, we obtain

$$P = \frac{\$20,000}{0.12} = \$166,667$$

These commonly used compound interest relationships are summarized in Table 5.1.

5.3 NOMINAL AND EFFECTIVE INTEREST RATES

It is often desirable to compound interest at more frequent intervals than yearly periods. For instance, a bank may pay interest on deposits quarterly, and will receive payments on loans monthly. In these instances, the period is a fraction of a year. When the compounding period is for other than one year, the interest rate will not represent an annual rate. An interest rate can be expressed as a *nominal annual rate* or *nominal rate*. The nominal rate is the rate per period times the number of periods per year. For example, an interest rate of 1 percent per period equal to a nominal annual rate of 1 percent if interest is compounded yearly, or a nominal annual rate of 12 percent if interest is compounded monthly.

One dollar invested at an *effective annual rate* of interest i grows in one year to $\$(1 + i)$ and in n years to $\$(1 + i)^n$ independent of how often the interest is compounded within a year. If the interest is compounded monthly at an effective monthly rate i_m, then in one year the dollar grows to $\$(1 + i_m)^{12}$ and in n years to $\$(1 + i_m)^{12n}$. The following relations exist between i and i_m:

$$1 + i = (1 + i_m)^{12} \tag{5.12}$$

TABLE 5.1
Discrete Interest Formulas

To Find Given	Factor	Name of Factor
$F = P$	$[(1 + i)^n]$	Single payment compound amount factor
$P = F$	$\left[\dfrac{1}{(1 + i)^n}\right]$	Single payment present worth factor
$A = F$	$\left[\dfrac{i}{(1 + i)^n - 1}\right]$	Sinking fund deposit factor
$A = P$	$\left[\dfrac{i(1 + i)^n}{(1 + i)^n - 1}\right]$	
$A = P$	$\left[\dfrac{i}{(1 + i)^n - 1} + i\right]$	Capital recovery factor (amortization factor)
$A = P$	$\left[\dfrac{i}{1 - (1 + i)^{-n}}\right]$	
$F = A$	$\left[\dfrac{(1 + i)^n - 1}{i}\right]$	Uniform series compound amount factor (final value of an annuity)
$P = A$	$\left[\dfrac{(1 + i)^n - 1}{i(1 + i)^n}\right]$	Uniform series present worth factor (present value of an annuity)

or

$$i = (1 + i_m)^{12} - 1 \qquad (5.13)$$

and

$$i_m = (1 + i)^{1/12} - 1 \qquad (5.14)$$

When a nominal interest rate is used, the nominal rate j is divided by the number of periods per year p, and the total number of periods is equal to the number of years times the periods per year. For instance,

$$\text{Single payment compound amount} = (\$1)\left(1 + \frac{j}{p}\right)^{pn} \qquad (5.15)$$

where

j = nominal yearly rate

n = number of years

p = number of interest conversion periods per year

j/p = uniform rate of interest per period

pn = number of interest conversion periods

The following relations exist between i and j,

$$(1 + i)^n = \left(1 + \frac{j}{p}\right)^{pn} \tag{5.16}$$

or

$$i = \left(1 + \frac{j}{p}\right)^{p} - 1 \tag{5.17}$$

and

$$j = p[(1 + i)^{1/p} - 1 \tag{5.18}$$

If interest is compounded *continuously*, p is equal to infinity, and from Eq. 5.16 the single payment compound amount becomes

$$(\$1)(1 + i)^n = \lim_{p \to \infty} \left(1 + \frac{j}{p}\right)^{pn} = (\$1)e^{jn} \tag{5.19}$$

$$\text{Present value} = (\$1)e^{-jn} \tag{5.20}$$

where e = 2.7183.
It follows that for continuous compounding,

$$\text{Effective rate} = i = e^j - 1 \tag{5.21}$$

and

$$\text{Nominal rate} = j = \ln (1 + i) \tag{5.22}$$

Thus, a *nominal annual rate* of 12 percent compounded continuously is equal to an *effective annual rate* of $e^{0.12} - 1 = 0.1275$ or 12.75 percent. The effective annual rate is the rate i compounded yearly that is equivalent to some nominal rate j. Table 5.2 shows the relationship between j and i (expressed as percentages) based on Eqs. 5.21 and 5.22, that is, assuming continuous compounding.

Equal amounts left to grow at compound interest at the same nominal rate, but compounded at a different number of periods per year, will vary considerably over a long period (see Table 5.3). Continuous compounding closely approximates monthly compounding. Thus continuous compounding, although physically impossible, may be used for its mathematical convenience. A plot of the relationship between i and j, over a wide range of values, is shown in Fig. 5.1.

TABLE 5.2
Nominal Interest Rate (j) Versus Effective Interest Rate (i) for Continuously Compounded Interest

$$j = \ln(1 + i) \qquad i = e^j - 1 \qquad (i \text{ and } j \text{ are fractions})$$

	$j \to i$					$i \to j$	
j (%)	i (%)	j (%)	i (%)	i (%)	j (%)	i (%)	j (%)
1	1.01	35	41.91	1	0.995	35	30.01
2	2.02	40	49.18	2	1.980	40	33.65
3	3.05	45	56.83	3	2.956	45	37.16
4	4.08	50	64.87	4	3.92	50	40.55
5	5.13	55	73.33	5	4.88	55	43.83
6	6.18	60	82.21	6	5.83	60	47.00
7	7.25	65	91.55	7	6.77	65	50.08
8	8.33	70	101.38	8	7.70	70	53.06
9	9.42	75	111.70	9	8.62	75	55.96
10	10.52	80	122.55	10	9.53	80	58.78
11	11.63	85	133.96	11	10.44	85	61.52
12	12.75	90	145.96	12	11.33	90	64.19
13	13.88	95	158.57	13	12.22	95	66.78
14	15.03	100	171.83	14	13.10	100	69.32
15	16.18	150	348.17	15	13.98	150	91.63
16	17.35	200	638.91	16	14.84	200	109.86
17	18.53			17	15.70		
18	19.72			18	16.55		
19	20.92			19	17.40		
20	22.14			20	18.23		
21	23.37			21	19.06		
22	24.61			22	19.89		
23	25.86			23	20.70		
24	27.12			24	21.51		
25	28.40			25	22.31		
26	29.69			26	23.11		
27	31.00			27	23.90		
28	32.31			28	24.69		
29	33.64			29	25.46		
30	34.99			30	26.24		

TABLE 5.3
Single Payment Compound Amount of $1 Left at 12 Percent Nominal Rate (j)

Year (n)	Period of Compounding			
	Yearly	Semiannually	Monthly	Continuously
1	1.1200	1.1236	1.1268	1.1275
5	1.7623	1.7908	1.8167	1.8221
10	3.1058	3.2071	3.3004	3.3201
20	9.6463	10.2857	10.8926	11.0232
50	289.0022	339.3021	291.5834	403.4288
100	83,522.2657	115,125.9039	153,337.5568	162,754.7914
Effective rate (i)	12%	12.36%	12.68%	12.75%

The distinction between nominal and effective rate is important. Continuous interest tables are often constructed for nominal rates j. Midyear and conventional (end of year tables) are often based on an effective rate i. Unfortunately, we rarely specify whether a rate is a nominal rate or an effective rate. This can result in a misunderstanding when we are dealing with another company that uses a different type of interest rate.

5.4 INCOME RECEIVED OTHER THAN AT THE END OF THE PERIOD

The equations presented so far are based on the assumption that income, or payment, is received at the end of a period. Income from producing oil and gas properties is usually received monthly or even continuously. Money received throughout a period has a greater value to the recipient than the same amount of money received at the end of the period, since it can be reinvested sooner. Several methods that simulate a monthly receipt of income but utilize yearly tables are commonly used in the petroleum industry.

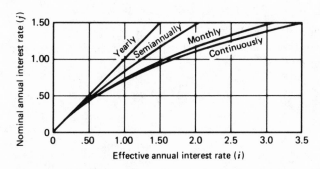

Figure 5.1   Relationship between effective rate i and nominal rate j for various compounding periods.

5.4.1   Midyear Factors

A commonly used method to account for monthly receipt of income is to assume that each year's income is received at the midpoint of the year. The following equations apply when midyear receipts of income are involved:

Single payment compound amount factor $= (1 + i)^{[n-(1/2)]}$

$$= (1 + i)^n(1 + i)^{-1/2} \tag{5.23}$$

Present value factor $= (1 + i)^{-[n-(1/2)]}$

$$= (1 + i)^{-n}(1 + i)^{1/2} \tag{5.24}$$

Uniform series present value factor $= \Sigma_1^n \, (1 + i)^{-[n-(1/2)]}$

$$= (1 + i)^{1/2} \left[\frac{1 - (1 + i)^{-n}}{i} \right] \tag{5.25}$$

Equations 5.23, 5.24, and 5.25 differ from Eqs. 5.1, 5.2, and 5.9 by the factor $(1 + i)^{\pm 1/2}$. Thus midyear interest formulas and conventional interest formulas differ only by a simple factor and are readily convertible. Both sets of formulas are based on an effective interest rate i.

5.4.2   Continuous Interest Formulas

Many people believe that it is more representative of actual business conditions to treat those transactions which occur fairly uniformly throughout the year as continuous cash flows. Because these cash flows are continuous and earnings are created uniformly throughout the year, interest is also treated continuously in such cases.

Instant Cash Flow and Continuous Compounding

Consider a fund of value P at $t = 0$, which increases to F at time $t = n$ as a result of continuous compounding at rate j. The increase in the fund due to interest earned during a time interval dt is

$$dF = jF \, dt \tag{5.26}$$

where

$$j = \frac{\$ \text{ interest/year}}{\$ \text{ in fund}} = \text{nominal continuous interest rate}$$

Thus

$$\int_P^F \frac{dF}{F} = \int_0^n j\, dt \tag{5.27}$$

$$\ln \frac{F}{P} = jn \tag{5.28}$$

and

$$\frac{F}{P} = e^{jn} \quad \text{or} \quad \frac{P}{F} = e^{-jn} \tag{5.29}$$

and

$$P = Fe^{-jn} \tag{5.30}$$

Hence e^{-jn} is the present value or discount factor relating a sum F, n years in the future, to a present sum P.

The other commonly used continuous interest factors for lump sum end of period cash flows are:

$$\text{Sinking fund factor} = \frac{A}{F} = \frac{e^j - 1}{e^{jn} - 1} \tag{5.31}$$

$$\text{Capital recovery factor} = \frac{A}{P} = \frac{(e^j - 1)e^{jn}}{e^{jn} - 1} \tag{5.32}$$

$$\text{Uniform series compound amount factor} = \frac{F}{A} = \frac{(e^{jn} - 1)}{e^j - 1} \tag{5.33}$$

$$\text{Uniform series present value factor} = \frac{P}{A} = \frac{e^{jn} - 1}{(e^j - 1)e^{jn}} \tag{5.34}$$

Continuous Cash Flow and Periodic Compounding

If an amount of money ($1/p$) is received and reinvested p times per year at some nominal rate j, then from Eq. 5.8 and for a single whole period, the value of the money is

$$F_1 = (1/p) \left[\frac{(1 + j/p)^p - 1}{j/p} \right]$$

$$= \frac{(1 + j/p)^p - 1}{j} = \frac{i}{j} \tag{5.35}$$

This is the relative increase of the money received and reinvested throughout the period over the value of the money if it were received at the end of the period. For continuous cash flow p tends to infinity and

$$F_1 = \frac{e^j - 1}{j} = \frac{i}{\ln (1 + i)}$$
(5.36)

Similarly, if an amount of $\$(1/p)$ is received and reinvested mp times, the value of the money reinvested after m years is

$$F_m = \left(\frac{1}{p}\right) \left[\frac{\left(1 + \dfrac{j}{p}\right)^{mp} - 1}{j/p} \right]$$
(5.37)

The relative increase in value of the money is

$$\frac{F_m}{m} = \frac{(1 + j/p)^{mp} - 1}{jm}$$
(5.38)

where m is a fractional part or multiple of a year. For continuous cash flow,

$$\frac{F_m}{m} = \frac{e^{jm} - 1}{jm} = \frac{(1 + i)^m - 1}{m \ln (1 + i)}$$
(5.39)

The factors in Eqs. 5.36 and 5.39 can be used to multiply the present value obtained for year-end cash flows to convert it to actual present value for continuous cash flow and periodic compounding. Applying Eq. 5.39 on Eq. 5.2, for example, we obtain

$$PV \text{ of } \$1 = (1 + i)^{-n} \left[\frac{(1 + i)^m - 1}{m \ln (1 + i)} \right]$$
(5.40)

Continuous Cash Flow and Continuous Compounding

If there is a continuous flow of money, $\$\bar{A}$ per year, over n years,

$$dF = \bar{A} \, dt$$
(5.41)

and the present value using continuous compounding is

$$dP = \bar{A} \, dt \, e^{-jt}$$
(5.42)

Thus

$$\int_0^P dP = \int_0^n \bar{A}e^{-jt}\, dt \tag{5.43}$$

or

$$P = \bar{A}\,\frac{(1 - e^{-jn})}{j} \tag{5.44}$$

Thus the uniform series present value factor is

$$\frac{P}{\bar{A}} = \frac{1 - e^{-jn}}{j} = \frac{e^{jn} - 1}{je^{jn}} \tag{5.45}$$

The funds flow term $\bar{A}$ is the single total amount of funds flowing continuously during a period, in a uniform series of n equal payments.

Rearranging Eq. 5.45 gives us

$$\frac{\bar{A}}{P} = \frac{je^{jn}}{e^{jn} - 1} \tag{5.46}$$

which is the capital recovery factor. To relate $\bar{A}$ and F, multiply Eq. 5.44 by e^{jn}:

$$Pe^{jn} = F = \frac{\bar{A}(1 - e^{-jn})e^{jn}}{j} \tag{5.47}$$

or

$$F = \frac{\bar{A}(e^{jn} - 1)}{j} \tag{5.48}$$

Thus, the uniform series compound amount factor is

$$\frac{F}{\bar{A}} = \frac{e^{jn} - 1}{j} \tag{5.49}$$

and the sinking fund factor is

$$\frac{\bar{A}}{F} = \frac{j}{e^{jn} - 1} \tag{5.50}$$

It may become necessary to find the present value of a funds flow single cash payment or receipt F^* made during the nth year. This can be achieved by combining the appropriate funds flow factor and end of period continuous interest factor. From Eq. 5.44, the sum of money at time $n - 1$ equivalent to F^* throughout period n is

$$P_{n-1} = \frac{F^*(1 - e^{-j})}{j} \tag{5.51}$$

The present value of this sum is (using Eq. 5.30)

$$P_0 = \frac{F^*(1 - e^{-j})}{j} e^{-j(n-1)} \tag{5.52}$$

or

$$P_0 = \frac{F^*(e^j - 1)}{je^{jn}} \tag{5.53}$$

The factor $(e^j - 1)/je^{jn}$ can be used to find the present value at time zero of one-year cash flow during the nth year. It should be noted that F^* is the single total amount of funds flowing continuously during a period in a single equivalent payment for the year occurring at the end of the funds flow payment period.

Example 5.7.   Find the present value of $1000 received at the end of the fifth year, if interest is compounded continuously at a nominal annual rate of 10 percent.

Solution

$$P = \$1000\, e^{-(0.10)(5)} = \$606.53$$

Example 5.8.   A sum of $10,000 is received uniformly over the fifth year at 20 percent annual nominal interest rate compounded continuously. What is the present value of this funds flow?

Solution
Using Eq. 5.53, we get

$$P = \$10,000 \left(\frac{e^{0.20} - 1}{0.20e^{1.0}} \right) = \$4072.48$$

Example 5.9.   A sum of $100,000 will be received uniformly over a five year period beginning five years from today. What is the present value of this funds flow if interest is compounded uniformly at a nominal annual rate of 10 percent?

Solution

Using Eq. 5.48 gives us

$$F_{10} = \frac{\bar{A}(e^{5j} - 1)}{j} = \frac{\bar{T}(e^{5j} - 1)}{5j}$$

where $\bar{T} = 5\bar{A}$ = total cash flow over the five year period. With Eq. 5.30,

$$P_0 = \frac{\bar{T}(e^{5j} - 1)e^{-jn}}{5j}$$

In general,

$$P = \bar{T}\left(\frac{e^{jm} - 1}{jm}\right)e^{-jn} \tag{5.54}$$

or

$$P = \bar{T}\left(\frac{1 - e^{-jm}}{jm}\right)e^{-j(n-m)} \tag{5.55}$$

Using the figures above gives us

$$P = \frac{\$100,000(e^{5(0.10)} - 1)e^{-(0.10)(10)}}{(5)(0.10)}$$

$$= \$47,730.24$$

Example 5.10. $5000 was invested on the purchase of land for a project two years before time zero for the project. What is the present value of this investment if the nominal annual continuous interest rate is 15 percent?

Solution

Using Eq. 5.30, we obtain

$$P_0 = F_n e^{-jn}$$

For negative n,

$$P_0 = F_{-n} e^{jn}$$

$$= \$5000 e^{(0.15)(2)} = \$6749.29$$

Example 5.11. $10,000 was invested uniformly from three years before to zero time at 15 percent nominal annual continuous interest rate. What is the present value of this investment at time zero?

Solution

From Eq. 5.48, from $t = n$ years to $t = 0$:

$$P = \frac{\bar{A}(e^{jn} - 1)}{j} = \frac{\bar{T}(e^{jn} - 1)}{jn}$$

where $T = n\bar{A}$ = total cash flow over n years.

$$P = \frac{\$10,000[e^{(0.15)(3)} - 1]}{(0.15)(3)} = \$12,629.16$$

Example 5.12.   Income from a project is projected to decline at a constant rate from an initial value F_0 at time zero to a final value F_n at time $t = n$. If interest is compounded continuously at a nominal annual rate j, determine the present value of the continuous cash flow.

Solution

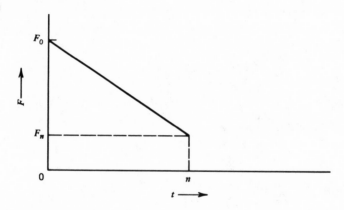

At any time t,

$$F = F_0 - \left(\frac{F_0 - F_n}{n}\right) t = F_0 \left(1 - \frac{t}{n}\right) + \frac{F_n}{n} t \qquad (5.56)$$

When discounted to time zero,

$$P = \int_0^n F e^{-jt} \, dt$$

or

$$P = F_0 \int_0^n \left(1 - \frac{t}{n}\right) e^{-jt} \, dt + \frac{F_n}{n} \int_0^n t e^{-jt} \, dt$$

$$= \frac{F_0}{j}\left(1 - \frac{1 - e^{-jn}}{jn}\right) + \frac{F_n}{nj}(1 - ne^{-jn} - e^{-jn}) \qquad (5.57)$$

For the special case where a total cash flow amounting to T declines from F_0 to zero in n years:

$$T = \frac{F_0 n}{2} \tag{5.58}$$

and

$$P = \frac{2T}{jn}\left(1 - \frac{1 - e^{-jn}}{jn}\right) \tag{5.59}$$

If $T = \$20,000$, $n = 10$ years and $j = 20$ percent,

$$P = \frac{(2)(\$20,000)}{(0.20)(10)}\left[1 - \frac{1 - e^{-(0.20)(10)}}{(0.20)(10)}\right]$$

$$= \$11,353.35$$

5.5   DISCOUNT FACTORS AND TABLES

From the foregoing discussion it is clear that in order to calculate the present value of an amount to be received at a future date, this amount has to be multiplied by a factor. This factor is called a *discount factor*, *present value factor*, or *deferment factor*, and is defined as follows:

A discount factor D *is the factor by which the cumulative (undiscounted) payments/receipts have to be multiplied to arrive at the present value of these payments/receipts. That is,*

$$D = \frac{\text{present value of future income/expense}}{\text{undiscounted value of future income/expense}}$$

or

$$\text{Present value} = (\text{discount factor}) \times (\text{future income/expense}) \tag{5.60}$$

Future income (expense) of an oil or gas field project may be (a) a fixed amount at a certain date, (b) a fixed amount at regular intervals, or (c) a declining amount at regular intervals. The present value of such payments/receipts may be obtained by multiplying the cumulative payments/receipts by the appropriate discount factor.

Figure 5.2 illustrates some common cash flows and their corresponding discount factors.

For single end of year payments and annual compounding, using Eq. 5.2 we obtain

$$D'_{sp} = \frac{F/(1 + i)^n}{F} = \frac{1}{(1 + i)^n} \tag{5.61}$$

Discount factors for annual compounding and year-end cash flows are given in Appendix A.

For single payments at the middle of the year and annual compounding, Eq. 5.24 yields

$$D''_{sp} = \frac{F/(1 + i)^{n-(1/2)}}{F} = \frac{1}{(1 + i)^{n-(1/2)}} \tag{5.62}$$

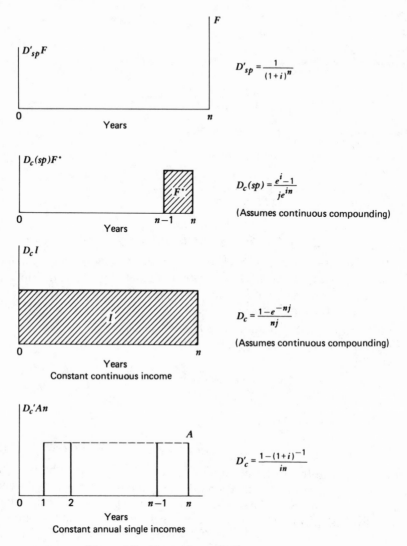

Figure 5.2   Illustration of Discount Factors.

Discount factors for annual compounding and midyear cash flows are given in Appendix B.

For single end of year payments and continuous compounding, Eq. 5.30 gives the discount factor as

$$D_{sp}''' = e^{-jn} \qquad\qquad (5.63)$$

Discount factors for continuous compounding and year-end cash flows may be found in Appendix C.

For multiple payments received at the end of the year and periodic compounding, Eq. 5.9 yields

$$D_c' = A \left[\frac{1 - (1 + i)^{-n}}{i} \right] / An$$

$$= \frac{1 - (1 + i)^{-n}}{in} \tag{5.64}$$

For multiple payments received at the middle of each year and periodic compounding,

$$D_c'' = A \left[\frac{1 - (1 + i)^{-n}}{i} \right] (1 + i)^{1/2} / An$$

$$= \frac{[1 - (1 + i)^{-n}](1 + i)^{1/2}}{in} \tag{5.65}$$

If interest is compounded continuously and income is received uniformly and continuously throughout the life of a project, the discount factor equation may be obtained from Eq. 5.44:

$$D_c = \frac{\bar{A}(1 - e^{-jn})}{j\bar{A}n} = \frac{1 - e^{-jn}}{jn} \tag{5.66}$$

For sums that occur uniformly over a period with continuous interest rates, the discount factor may be obtained from Eq. 5.53:

$$D_c''' = \frac{e^j - 1}{je^{jn}} \tag{5.67}$$

Discount factors for sums that occur uniformly over one-year periods after time zero are given in Appendix D.

The most appropriate discount (interest) rate used for present value calculations is often the rate of interest to be paid on borrowed capital, that is, the cost of investment capital. All other considerations, such as risk, inflation, and interest paid, should be included internal to the cash flow calculations and not allowed to affect the discount rate.

REFERENCES

Essley, P. L., Jr.: "The Difference Between Nominal and Effective Interest Tables and Nominal and Effective Rates of Return," *Journal of Petroleum Technology*, August 1965, pp. 911–918.

McCray, A. W.: *Petroleum Evaluations and Economic Decisions*, Prentice-Hall, Englewood Cliffs, New Jersey, 1975.

Newendorp, P. D.: *Decision Analysis for Petroleum Exploration*, Petroleum Publishing Company, Tulsa, 1975.

Stermole, F. J.: *Economic Evaluation and Investment Decision Methods*, Third Edition, Investment Evaluations Corporation, Golden, Colorado, 1980.

Uhl, V. W. and Hawkins, A. W.: *Technical Economics for Engineers*, American Institute of Chemical Engineers Continuing Education Series 5, American Institute of Chemical Engineers, New York, 1971.

6

PROFITABILITY OF
A VENTURE

6.1 INTRODUCTION

Companies usually consider the possible benefits they may derive from ventures before investing money. Making sound capital expenditure decisions requires an objective means of measuring the productivity of individual investment proposals. Since the best measure of the economic worth of investment proposals is their ability to produce profits, it is common practice to grade proposed expenditures in terms of an objective, a means of evaluating or measuring proposed investments and financial opportunities, and a criterion for their acceptance or rejection.

The financial benefits, expressed by the *profitability* of the investments, are measured by profitability criteria, profit indications, or economic yardsticks. In developing yardsticks, it must be realized that in order to decide if a particular investment should be made, it should be compared with other investment opportunities. This means that, even if this particular investment is considered to be profitable, if the money can be spent to greater advantage in some other endeavor, it may be decided that this investment does not meet the proper profitability qualifications. There are basically two kinds of yardsticks: *screening* and *ranking* yardsticks. A screening yardstick determines which ventures meet the minimum qualifications to be considered for investment; a ranking yardstick determines which of two or more mutually exclusive ventures is the most desirable from the point of profitability.

The time value of money has long been recognized, and the search for improved profitability criteria has been continuous. However, time-related profitability criteria did not gain wide acceptance in the oil and gas industry until the late 1950s. During the post World War II development drilling boom, oil companies were earning some 20 percent net returns on net shareholder investments; money was available at 3 and 4 percent interest rates. Management perhaps did not feel a need for highly refined profitability criteria.

However, in the late 1950s profit margins began to shrink, drilling and development prospects were harder to find and less profitable when found and, as a direct consequence, there arose a need for better profitability criteria. An article by Dean on the discounted cash flow *(DCF)* rate of return provided the vehicle for the acceptance of this profitability criterion in the oil and gas industry. In the 1960s numerous articles like the one by Jackson appeared dealing with the cost of capital and suggesting that profit discounted at the cost of capital is the most reliable criterion. This seems to be the profitability criterion preferred by the professionals in the field today.

Since the 1960s major advances have been made in profitability analysis. Attempts have been made to quantify risk and uncertainty. New theories and procedures known as decision theory, preference theory, or utility theory have emerged. The acceptance of these advances by decision makers has been quite gradual.

In the following discussion a brief review is given of some of the most used profitability criteria. No single criterion tells us everything about profitability. So we normally choose the criterion (or suite of criteria) that gives the explorationist and management the maximum amount of information about economic factors.

A good profitability criterion should have the following characteristics (Newendorp). It should:

1. Be suitable for comparing and ranking the profitability of various investment opportunities.

2. Reflect the firm's "time value" of capital.

3. Indicate whether profitability exceeds a minimum, such as the cost of capital and/or the firm's average earning rate.

4. Include quantitative statements of risk (probability numbers), if possible.

5. Reflect other factors such as corporate goals, decision makers risk preferences, and firm's asset position, if possible. (This involves utility or preference theory.)

6.2   PAYOUT (OR PAYBACK) PERIOD

The payout period is the oldest and simplest profit indicator. It is the time required for the cumulative net earnings to equal the initial outlay; the length of time required to get our investment capital back. The payout period thus measures the speed with which invested funds are returned to the business.

The shorter this period is found to be, the higher the project is rated. On the other hand, if the calculated period is deemed too long, the project is rejected. A survey of American practice indicates that a maximum allowable payout period is commonly set at 3 to 5 years; rarely as high as 10 years.

The speed with which invested funds are returned to the business may be a criterion of some importance in capital-investment decision making under two circumstances:

1. When the funds available for capital investment are limited compared with the opportunities that exist for profitable investment and when it is expected that future investment opportunities will be more profitable than the current ones.

2. When a high degree of uncertainty exists as to how long the profitability of a proposed venture will continue into the future.

When one or more of the above conditions is present, the use of payout period as *one* of the investment criteria may be justified.

Because it is relatively simple to calculate and because managers instinctively like to recover their investment as rapidly as possible, the payout calculation is frequently used to evaluate investment proposals. When used in conjunction with other measures, the payout period is useful in promoting wise investment decision making. When used as a sole or principal criterion for investment decisions, the payout period is dangerous because it may result in the choice of less profitable investments that yield high initial returns for short periods as compared with more profitable investments that provide profits for longer periods of time.

The main shortcomings of this yardstick are as follows:

1. It does not consider the income to be obtained after the payout period.

2. It ignores the timing of the returns that occur prior to and after the payout date.

3. It only has meaning if no substantial investments are made after the initial investment.

4. It obviously fails as a ranking yardstick.

Example 6.1. The following data are used to illustrate calculations of some of the profitability criteria discussed.

Drilling costs: $1,500,000

Future series of cash flow (revenues) generated by the investment:

Year	Cash Flow (Revenue)
1	$1,000,000
2	800,000
3	600,000
4	400,000
5	200,000
6	100,000
	$3,100,000

Cash flows are after tax.

The end of year convention is used. Revenues are assumed to be received at year-end; annual compounding is also assumed.

The average opportunity rate is assumed to be 15% per year.

Using the above data, calculate the payout period for the hypothetical prospect.

Solution

Investment: $1,500,000

Unrecovered portion of investment after the first year:

$$\$1,500,000 - \$1,000,000 = \$500,000$$

Fraction of second year required to recover this remaining balance:

$$\frac{\$500,000}{\$800,000} = 0.625$$

Therefore, the payout period for our prospect = 1.625 years.

Example 6.2. Data for two alternative projects A and B are given below. Select one of the projects based on the payout criterion.

Solution

	Project A	Project B
Investment	$250,000	$250,000
Annual income	50,000	75,000
Payout (years)	$\dfrac{250,000}{50,000} = 5$	$\dfrac{250,000}{75,000} = 3.33$

Project B, with a shorter payout period, would be the better investment proposal.

6.3 PROFIT-TO-INVESTMENT RATIO

One weakness of payout time as a measure of project profitability is that it does not consider the total profit from the investment project. The profit-to-investment ratio reflects total profitability. It is defined as the ratio of total net (undiscounted) profit to the investment. It is a dimensionless number relating the amount of new profit generated per dollar invested. Let C be the initial investment and I the net income derived from it. Then the profit-to-investment ratio

$$P/\$ \text{ invested} = \frac{I - C}{C} \tag{6.1}$$

Example 6.3. Using the data of Example 6.1, calculate the $P/\$$ invested.

Solution

$$I = \$3,100,000 = \text{total net revenue}$$

$$C = \$1,500,000 = \text{investment}$$

$$\text{Total net profit} = \$3,100,000 - \$1,500,000 = \$1,600,000$$

$$\text{Profit-to-investment ratio:} \frac{\$1,600,000}{\$1,500,000} = 1.067 = 106.7\%$$

Some companies use the net income-to-investment ratio. Using the data above, we get

$$\text{Net income-to-investment ratio} = \frac{\$3,100,000}{\$1,500,000} = 2.067 \quad \text{or} \quad 206.7\%$$

The profit-to-investment ratio is easy to calculate and can be expressed in before tax or after tax values. The objective would be to select prospects that maximize profits per unit of money invested. The major weakness of this ratio is that it does not reflect the time-rate pattern of income from the project.

A Combination of Pay Out and Profit Per Dollar Invested

This has often been used as a screening criterion. If a venture will yield a $P/\$$ larger than a minimum value, which is obtained with a payout period less than a certain maximum value, then it may be considered for investment. This screening can be quite useful where investments in ventures of similar nature are compared, such as drilling of oil and gas wells.

6.4 NET PRESENT VALUE PROFIT

The net present value *(NPV)* or net present worth profit is the algebraic sum of all net cash flows when discounted to time zero using a single, previously specified discounting rate. The decision rule is to accept projects that maximize *NPV* profit and reject all projects having negative *NPV* profit.

The greater the positive *NPV* for a project, the more economically attractive it is. A project with a negative *NPV* is not a profitable proposition. If, however, a project is not concerned with making a profit, but with meeting a necessary objective at the minimum overall cost, for example, investment in pollution control to meet required standards, its *NPV* will be negative. The economic aim in considering alternatives is to reduce the negative *NPV* (net present cost) to as small a level as possible.

Example 6.4. Using an average reinvestment opportunity rate of 15 percent, calculate the net present value profit of the cash flow of Example 6.1.

Solution

Year	Net Cash Flow	Discount Factor $i^* = 0.15$ (Year-End)	Discounted Cash Flow
0	− $1,500,000	1.0000	− $1,500,000
1	+ 1,000,000	0.8696	+ 869,960
2	+ 800,000	0.7561	+ 604,880
3	+ 600,000	0.6575	+ 394,500
4	+ 400,000	0.5718	+ 228,720
5	+ 200,000	0.4972	+ 99,440
6	+ 100,000	0.4323	+ 43,230
		$NPV_{i^*=0.15}$ =	+ $ 740,730

Net present value discounted at 15 percent is positive. This means that the six annual cash revenues are preferred to our initial investment of $1,500,000 if the discount rate is 15 percent. If we invest the $1,500,000 we would make a 15 percent rate of return plus increase our net worth by $740,730.

The *NPV* takes account of all earnings throughout the expected life of the asset and is simpler to calculate than the rate of return (discussed later) because the trial-and-error solutions are eliminated. When comparing alternatives with different expected lives, an explicit assumption is made in the evaluation of the future rates that a company's fund can earn. This compares with the implicit assumption made in the rate-of-return approach that monies can be reinvested at the same rates as monies earned by the projects under consideration.

However, for this criterion it remains a problem to determine the minimum acceptable rate of return that projects are expected to earn. What is the minimum rate of return that a proposal should earn to justify investment of the business's funds?

Except where other intangible benefits are involved, this rate of return should certainly be above the cost of capital to the firm. If the rate of return on a proposed investment is not above the cost of the funds, the firm is better off not investing its funds.

The acceptable rate of return for new investments should be above the cost of capital to the firm by an amount that will cover investments that must be made even though they will not earn an adequate return. Examples of this include investments for facilities to promote employee morale or safety.

As a practical matter, the acceptable rate of return should rarely approach the cost of capital. Effective management is usually capable of developing investment opportunities at more attractive rates of return that exceed the available finances of the company and that exceed the available time for management to cope with the potential new projects.

The problem of determining the firm's cost of capital is complicated by several factors: the company's funds usually come from various sources, such as preferred and common stockholders, bondholders, and retained earnings. The cost of money raised by issuing bonds or stock fluctuates markedly from time to time, and for capital-planning purposes we are interested in the relatively long-term pattern of future costs.

We must, therefore, estimate the relative proportions of future funds that will come from these various sources, the future costs of funds from each of these sources, and then compute a weighted average of these estimates.

Cost of Capital

The cost of capital is the weighted-average cost of investment capital, expressed as a percentage, from all sources. Prudent practice dictates that a company not invest in projects returning less than its cost of capital. Therefore, the minimum discount rate for present value calculations is the cost of capital.

Example 6.5.
(a) A company took a bank loan at 10 percent interest for a project. What is the cost of capital?
(b) Investment money for a proposed project is obtained from three lending agencies as follows:

> Bank A: $2,000,000 at 12%
> Bank B: $4,000,000 at 13%
> Bank C: $6,000,000 at 15%

What is the cost of capital?

Solution
(a) Before tax cost of capital = 10%
Using a 50% corporate tax rate, after tax cost of capital = $10(1 - 0.5) = 5\%$
(b) Total amount = $12,000,000
Before tax cost of capital = $[2(12) + 4(13) + 6(15)]/12 = 13.83\%$
After tax cost of capital = $(13.83)(1 - 0.5) = 6.92\%$

The cost of capital is usually greater than the cost of borrowed money. When a company sells bonds at 5 percent interest to raise capital, it changes the capital structure of the company and increases the risks of the equity (stockholder) capital. Should business conditions become unfavorable, the equity capital must sustain any losses before debt capital is impaired.

When the ratio of debt to equity gets high, the cost of equity capital rises. The high leverage makes the equity in the business more risky and investors expect a higher return on their equity investments to compensate for the higher risks. Other things being equal, the rate of interest that must be paid for the borrowed funds increases as the proportion of debt to equity capital in the company increases. For this reason alone, it is not logical to consider the cost of borrowed money as the cost of capital. The cost is usually considerably greater. Projects should earn substantially more than the cost of borrowed capital to justify investment of capital funds.

Thus the minimum acceptable rate of return to use in the net present value calculations is usually set by top management after consideration of at least some of the following factors:

1. Future investment opportunities and their anticipated rates of earnings.

2. If investment capital is borrowed, i^* must at least exceed the interest rate of the loan, or should at least exceed the average cost of capital.

3. Corporate growth objectives (the rate at which management has set for annual growth rate of treasury) should be taken into account.

6.5 DISCOUNTED CASH FLOW RATE OF RETURN (DCFROR)

DCFROR is also known by other names such as *internal rate of return, internal yield, profitability index*, and *interest rate of return*. It is defined as the discount rate that makes the *NPV* of a project equal to zero. Thus,

$$NPV = \sum_0^n \frac{A_n}{(1 + i)^n} = 0 \qquad (6.2)$$

or, in the continuous form,

$$NPV = \int_0^n A(t)e^{-jn} \, dt = 0 \qquad (6.3)$$

The discounted cash flow method is based on the principle that in making an investment outlay, we are actually buying a series of future annual incomes. The mechanics consist of finding the interest rate that discounts future earnings of a project down to a present value equal to the project cost. This interest rate is the rate of return on that investment.

DCFROR is computed by a trial-and-error series of calculations, as summarized below:

1. List the annual cash flows.

2. Select a discount rate and list discount factors. It is important at this point to identify end of year, midyear, and continuous factors correctly.

3. Calculate the present value of each annual cash flow and add the discounted values to obtain the net present value of the cash flow.

4. If the *NPV* is positive, select a higher discount rate. If the *NPV* is negative, select a lower discount rate.

5. After several trials and the zero *NPV* is bracketed, interpolate to find the *DCFROR*.

Example 6.6. Calculate the *DCFROR* for the cash flow of Example 6.1.

Solution
The final stages of the trial and error computation are as follows:

	First Trial, $i = 0.35$			Second Trial, $i = 0.45$	
Year	Cash Flow (Year-End)	Discount Factor ($i = 0.35$)	Discounted Cash Flow	Discount Factor ($i = 0.45$)	Discounted Cash Flow
0	−$1,500,000	1.0000	−$1,500,000	1.0000	−$1,500,000
1	+ 1,000,000	0.7407	740,740	0.6897	689,655
2	+ 800,000	0.5487	438,957	0.4756	380,499
3	+ 600,000	0.4064	243,865	0.3280	196,810
4	+ 400,000	0.3011	120,427	0.2262	90,487
5	+ 200,000	0.2230	44,603	0.1560	31,203
6	+ 100,000	0.1652	16,520	0.1076	10,759
			+$ 105,112		−$ 100,587

Interpolating between 0.35 and 0.45,

$$\text{Rate of return} = 0.35 + \left(\frac{105{,}112}{105{,}112 + 100{,}587} \right) (0.10)$$

$$= 0.4011 \quad \text{or} \quad 40.11\%$$

Investing $1,500,000 to buy the future series of six annual revenues is equivalent to investing $1,500,000 in a project that pays 40.11 percent compound annual interest.

Present Value Profile

Much of the confusion that results from the use of profitability criteria can be eliminated by plotting the present value profit versus the discount rate. This curve is called the *present value profile*. The present value profile for the prospect considered in Example 6.6 is shown in Fig. 6.1. This figure also illustrates the concept of rate of return, the point where the present value profile crosses the discount rate axis.

From Fig. 6.1 the net profit of $1,600,000 occurs at the zero discount axis. The intersection of the profile with the discount rate axis gives a discounted cash flow rate of return of about 40 percent. For discount rates less than 40 percent the discounted net present value is positive (for example, $741,000 at 15 percent). This indicates that the prospect should be accepted if the cost of capital is less than 40 percent. For discount rates greater than 40 percent, the discounted net present value profit is negative indicating rejection of the prospect. Changes in the initial investment simply shift the profile in the vertical direction by the amount of this change.

Example 6.7.   Compare the profitability of the following two investment proposals:

Proposal A: An investment of $100,000 today to receive $120,000 continuously in one year.

Proposal B: An investment of $100,000 today to receive $200,000 continuously in seven years.

Use continuous discounting.

Solution
The present value profiles for investment proposals A and B are shown on Fig. 6.2. Proposal A has a net profit of $20,000 as indicated on the zero discount rate axis, a profit-to-investment ratio of 0.2, and a discounted cash flow rate of return of 37.5 percent. By comparison, proposal B has net profit of $100,000, a profit-to-investment ratio of 1.0, and a discounted cash flow rate of return of 22.8 percent.

Data for preparation of the present value profile are shown below. The applicable formula is Eq. 5.44 of Chapter 5.

Discount Rate j%	NPV Proposal A ($)	NPV Proposal B ($)
0	20,000	100,000
5	17,049	68,750
10	14,195	45,833
15	11,434	23,821
20	8,762	7,629
25	6,176	(5,574)
30	3,673	(16,424)
35	1,250	(25,412)
40	(1,096)	(32,915)
45	(3,368)	(39,229)
50	(5,567)	(44,583)

This example brings out the weakness of the profit-to-investment ratio as a yard stick. Using this criterion, proposal B is a better investment proposal than A. However, the profit-to-investment ratio does not reflect the time-rate pattern of income from the prospects, and proposal B may not necessarily be better than proposal A.

The discounted cash flow rate of return, on the other hand, indicates A to be the better investment proposal, while the net present value at 15 percent indicates B to be the better investment proposal. The present value profiles give the whole picture. The profiles intersect at a discount rate of 19.5 percent. This is the break-even point. Investment proposal B gives higher relative profit if the cost of capital is less than 19.5 percent, and A is the better investment proposal if the cost of capital is higher than 19.5 percent.

The *DCF* rate of return has many advantages as a criterion for evaluating the desirability of investments. The concept of rate of return as a measure of profit-

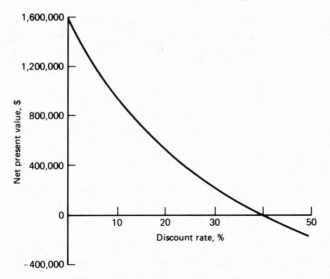

Figure 6.1 Present value profile.

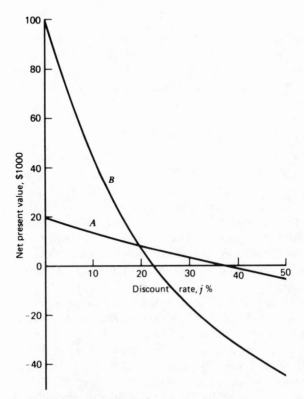

Figure 6.2 Present value profiles for proposals *A* and *B*.

ability is simple to understand and is directly related to the profit goals of the enterprise. Because profitability is stated as a percentage, it facilitates comparisons of alternatives involving different total sums. This criterion enables management to select for approval those investments showing the highest rates of return. It takes into account the time-rate patterns as to when cash flows are received. It is a convenient measure of value to compare with a "minimum" such as the cost of capital, or a corporate objective of a percentage annual growth rate. It introduces "time-value of money" concept into profitability analysis.

The *DCF* rate of return, however, has the following disadvantages and weaknesses as a criterion for evaluating investment opportunities:

1. The *DCF* rate of return is independent of the magnitude of the investment. The amount of investment capital required is important in determining whether or not we can embark on the project.

2. Some cash flow streams discount to zero for more than one interest rate. These cash flow streams are said to have multiple rates of return. Selection among investment proposals in such cases becomes quite difficult if *DCF* rate of return is used as the only criterion.

3. It may not be always possible to calculate a *DCF* rate of return. This occurs when cash flows are all negative or all positive, or when total undiscounted revenue is less than the investment.

4. The principle of the *DCF* rate of return involves the assumption that all cash flows are reinvested the instant they are received at the computed rate of return. This reinvestment requirement is the main weakness of the rate of return. It is not often that we are able to reinvest the income at the high rates of return such as are calculated.

Net Present Value Profit Versus *DCF* Rate of Return

Although *DCF* rate of return is widely used in project evaluation, it is more limited than *NPV* in its application. *NPV* gives a direct cash measure of a project's attractiveness. *NPV*s are additive when dealing with multiple project selection, an essential requirement in many cases. *NPV* can be used for all types of cash flows—even those in which terms are all negative. *NPV* is completely compatible with risk analysis techniques. *DCFROR*s on the other hand are rates of return on varying levels of investment and are not additive. *DCFROR* is a measure of the efficiency with which capital is employed and indicates the earning power of the project investment.

A common case in point illustrates the restriction in application of *DCFROR* and explains why there is a trend away from *DCFROR* toward *NPV*. Where a choice is to be made between alternative proposals for a project, these proposals have different cash flow patterns and different investments. In order to maximize the magnitude of the project's profits, the alternative with the greatest *NPV* should be selected. However, this is *not* necessarily the same as selecting the alternative with the highest *DCFROR*. The reason is that the investment upon which the return is measured is not constant. A lower return on a larger investment could yield a greater cash profit than a higher return on a smaller investment. In order to select the best alternative using *DCFROR* instead of *NPV*, it is necessary to look at the *DCFROR*s of the incremental cash flows between alternatives.

6.6   DISCOUNTED PROFIT-TO-INVESTMENT RATIO (DPR)

The discounted profit-to-investment ratio is the ratio of *NPV* profit discounted at i^* to the *NPV* of investment. This ratio is sometimes called *investment efficiency, present value index (PVI)* or *discounted profit ratio (DPR)*. Since the present value is dependent upon the assumed discount rate, the ratio must always be specified with its discount rate.

   DPR is a modification of *NPV* that is used to select projects under conditions of limited capital. It ranks projects properly to assure maximization of profits. The decision rule is to maximize *DPR*. The ratio will also become negative for discount rates greater than the calculated *DCFROR* of the project. All projects with negative *DPR*s are rejected outright.

Example 6.8.   Calculate the *DPR* for the cash flow of Example 6.1.

Solution

$$NPV \text{ at } i^* = 0.15 = \$740,730$$

$$DPR = \frac{\$740,730}{\$1,500,000} = 0.494$$

   The discounted profit-to-investment ratio includes all advantages of *NPV* in addition to providing a measure of profitability per dollar invested. It reflects the time-rate pattern of income from the project and thus is much more useful than the profit-to-investment ratio as a profitability criterion.

   The discount rate used to calculate *DPR* is usually low and may be: (a) the bank interest rate or company cost of money, (b) the average growth rate of company, or (c) the average rate of return of all company investments.

   DPR is a suitable measure of value for ranking and comparing investment opportunities. To maximize cumulative *NPV* for nonmutually exclusive alternatives, select the project with the largest *DPR* first, next largest second, and so on. It should be noted that *DPR* does not rank mutually exclusive projects properly. Incremental *DPR* analysis should be used to evaluate mutually exclusive alternatives.

   The use of discounted profit-to-investment ratio is increasing rapidly because of its utility. It is responsive both to timing and amount of profit, and it gives the profit per invested dollar in excess of that which is available through average company investment or by safe investment. The *DPR* is a measure of a project's contribution to company growth, and some authorities have suggested its sole use in ranking projects.

NPV profit and discounted profit-to-investment ratio are based on essentially the same logic as regards the time value of money concept. Maximizing *NPV* profit is a sufficient decision criterion if the firm has more money than available investment projects. If there are more projects than money (a limited capital constraint), then management should accept projects that maximize the discounted profit-to-investment ratio.

6.7   GROWTH RATE OF RETURN (GRR)

The concept growth rate of return *(GRR)* is illustrated in Fig. 6.3, which shows the net after tax cash flow resulting from a project. To calculate *GRR*, we first compound all the positive cash flows forward to some time horizon, *t*, years in the future. Any cash flows beyond that time are discounted back to that point. The rate at which we compound (or discount) these cash flows is the reinvestment rate or opportunity rate for the company. Let the sum of these positive cash flows compounded (or discounted) to time *t* be *B*. We then discount the negative cash flows (investments) to the present at the same rate to get a total discounted investment, *I*.

Thus our project promises to give us *B* dollars at time *t* if we invest the equivalent of *I* dollars now. The rate of return of this investment proposal is the project's *GRR*. For annual compounding,

$$I(1 + GRR)^t = B \tag{6.4}$$

or

$$GRR = (B/I)^{1/t} - 1 \tag{6.5}$$

For continuous compounding,

$$Ie^{GRR \cdot t} = B \tag{6.6}$$

or

$$GRR = \frac{1}{t} \ln\left(\frac{B}{I}\right) \tag{6.7}$$

where

I = discounted investment

t = number of years to time horizon

B = compounded value (at the time horizon) of the positive cash flows

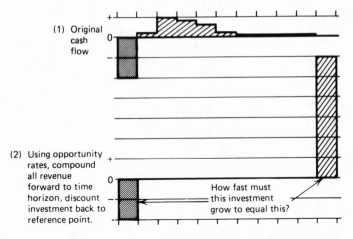

(1) Original cash flow

(2) Using opportunity rates, compound all revenue forward to time horizon, discount investment back to reference point.

How fast must this investment grow to equal this?

Figure 6.3 Concept of growth rate. (courtesy of SPE of AIME).

Let us show how the growth rate of return *(GRR)* is related to the net present value *(NPV)*. The net present value may be written for annual compounding and discounting as

$$NPV = B(1 + i)^{-t} - I \qquad (6.8)$$

where i is the reinvestment rate.

Equation 6.8 may be written as

$$B = (NPV + I)(1 + i)^t \qquad (6.9)$$

Substituting Eq. 6.9 in Eq. 6.4 gives us

$$I(1 + GRR)^t = (NPV + I)(1 + i)^t \qquad (6.10)$$

or

$$GRR = \left(\frac{NPV}{I} + 1\right)^{1/t}(1 + i) - 1 \qquad (6.11)$$

For continuous compounding and discounting

$$NPV = Be^{-jt} - I \qquad (6.12)$$

or

$$B = (NPV + I)e^{jt} \qquad (6.13)$$

Substituting Eq. 6.13 in Eq. 6.6, we get

$$Ie^{GRR \cdot t} = (NPV + I)e^{jt} \tag{6.14}$$

or

$$GRR = \frac{1}{t} \ln \left(\frac{NPV}{I} + 1 \right) + j \tag{6.15}$$

Equations 6.11 may be written as

$$\frac{1 + GRR}{1 + i} = \left(\frac{NPV}{I} + 1 \right)^{1/t} \tag{6.16}$$

Thus, if *NPV* is negative, *GRR* < *i*; and if *NPV* is positive, *GRR* > *i*. The same relationship exists between *GRR* and *j* for continuous compounding and discounting. Thus, *GRR* is equivalent to *NPV* as an accept/reject criterion. Any project for which the growth rate of return is greater than the opportunity rate.

Equations 6.11 and 6.15 show a definite relationship between *GRR* and discounted profit-to-investment ratio *(DPR)*. It is evident that if a project has a higher *DPR (NPV/I)* than another, it also has a higher *GRR*. Growth rate of return thus gives the same ranking of projects as *DPR*.

The growth rate of return, as a profitability criterion, yields the same accept/reject decisions as net present value and the same ranking of projects as discounted profit-to-investment ratio. Its calculations are straightforward, requiring no trial-and-error methods and yielding no multiple solutions as *DCFROR*. In addition, it is expressed as a rate of return.

Capital Allocation

Few people have more money than projects available. Managers are thus faced with the problem of choosing that set of projects that does not exceed the budget but maximizes the worth of the company over time.

Let us consider 11 projects with greatly differing cash flow patterns as given by Capen et al. These projects are shown in Table 6.1. Each project requires only one investment made at time zero; and each project has a 10-year life. Profit indicators like profit-to-investment ratio, payout, discounted cash flow rate of return, net present value (three discount rates), growth rate of return (three discount rates), and discounted profit-to-investment ratio (three discount rates). We have only $3000 and must choose among these 11 projects.

In Table 6.2 the projects have been ranked according to the profitability criteria. These show conflicts among the profit indicators. How shall we decide which projects are the best for our money? We would definitely be interested in the projects that generate the most value over time.

Consider, for example, Project Y. We invest $1000 at time zero and receive proceeds after tax of $210 each year. If the proceeds are reinvested at 10 percent,

TABLE 6.1
Project Comparisons: After Tax Net Cash Flow by Year (in Dollars)

Year	P	Q	R	S	T	U	V	W	X	Y	Z
0	−1,000	−1,000	−1,000	−1,000	−1,000	−1,000	−1,000	−500	−1,000	−1,000	−500
1	500	600	25	10	800	700	0	0	900	210	0
2	400	400	50	20	300	500	0	0	200	210	10
3	300	300	50	50	100	200	100	25	100	210	15
4	200	200	100	100	50	100	100	25	50	210	25
5	100	50	200	200	50	50	100	50	50	210	50
6	50	20	300	300	20	10	300	150	10	210	150
7	40	10	400	400	20	10	500	250	10	210	250
8	30	10	500	500	10	10	500	350	0	210	350
9	20	5	500	500	10	10	500	350	0	210	350
10	10	5	100	400	10	10	500	250	0	210	300

Criteria

	P	Q	R	S	T	U	V	W	X	Y	Z
Profit-to-investment ratio	0.65	0.60	1.23	1.48	0.37	0.60	1.60	1.90	0.33	1.10	2.00
Payout, years	2.3	2.0	6.7	6.8	1.7	1.6	6.8	7.0	1.5	4.8	7.0
DCFROR, %	22.5	25.0	12.5	13.4	19.3	28.5	13.8	14.7	20.0	16.5	15.1
Net present value											
at 5%, $	446	432	582	724	249	448	794	487	227	622	518
at 10%, $	284	294	153	230	150	323	270	188	139	290	208
at 15%, $	153	179	−141	−102	66	217	−81	−10	64	54	3
Growth rate											
at 5%, %	8.9	8.8	9.9	10.9	7.4	9.0	11.3	12.4	7.2	10.2	12.7
at 10%, %	12.8	12.9	11.6	12.3	11.5	13.1	12.7	13.6	11.4	12.8	13.9
at 15%, %	16.7	16.9	13.3	13.8	15.7	17.3	14.0	14.8	15.7	15.6	15.1
Discounted profit-to-investment (NPV per dollar invested)											
at 5%	0.45	0.43	0.58	0.72	0.25	0.45	0.79	0.97	0.23	0.62	1.04
at 10%	0.28	0.29	0.15	0.23	0.15	0.32	0.27	0.38	0.14	0.29	0.42
at 15%	0.15	0.18	−0.15	−0.10	0.07	0.22	−0.08	−0.02	0.06	0.05	0.01

Source: Capen et al.

TABLE 6.2
Project Rankings for Capital Allocation

Measure	Project										
	P	Q	R	S	T	U	V	W	X	Y	Z
Profit/investment	7	8	5	4	10	9	3	2	11	6	1
Payout	5	4	7	8	3	2	9	10	1	6	11
DCFROR	3	2	11	10	5	1	9	8	4	6	7
Net present value											
at 5%	8	9	4	2	10	7	1	6	11	3	5
at 10%	4	2	9	6	10	1	5	8	11	3	7
at 15%	3	2	11	10	4	1	9	8	5	6	7
Growth rate											
at 5%	8	9	6	4	10	7	3	2	11	5	1
at 10%	6	4	9	8	10	3	7	2	11	5	1
at 15%	3	2	11	10	4	1	9	8	6	5	7
Discounted profit-to-investment ratio											
at 5%	8	9	6	4	10	7	3	2	11	5	1
at 10%	6	4	9	8	10	3	7	2	11	5	1
at 15%	3	2	11	10	4	1	9	8	6	5	7

Source: Capen et al.

at the tenth year Project Y will provide a sum of $3346. The future value per dollar invested is $3.346. The future values and future value per dollar invested for all the projects are shown below. We can go down the list choosing those projects that give the largest future value in 10 years per dollar invested. Our $3000 will buy four projects Z, W, U, and Q, in order of their attractiveness. This ranking turns out to be identical with the ranking using growth rate of return or discounted profit-to-investment ratio.

Future Values of Projects

Project	Future Value ($)	Future Value per Dollar Invested ($)
P	3,331	3.331
Q	3,356	3.356
R	2,989	2.989
S	3,190	3.190
T	2,982	2.982
U	3,431	3.431
V	3,293	3.293
W	1.784	3.568
X	2,955	2.955
Y	3,346	3.346
Z	1,836	3.672

Source: Caper et al.

The list below shows the ranking by the other profit indicators, and the future value of money that the $3000 will purchase at the tenth year. *GRR* identifies projects that give the most money at the tenth year.

Criterion	Projects in Order of Choice	Future Value Money at the Tenth Year ($)
Profit/investment	Z, W, V, S	10,103
Payout	X, U, T	9,368
DCFROR	U, Q, P	10,119
Present value at 10%	U, Q, Y	10,134
Growth rate at 10%	Z, W, U, Q	10,408

Source: Capen et al.

REFERENCES

Capen, E. C., Clapp, R. V., and Phelps, W. W.: "Growth Rate—A Rate-of-Return Measure of Investment Efficiency," SPE Reprint Series No. 16 Economics and Finance, 1982, pp. 115–127.

Jackson, E. G., Jr.: "Here's a Way to Figure Your Cost of Capital," *Oil and Gas Journal*, January 9, 1967.

Newendorp, P. D.: *Decision Analysis for Petroleum Exploration*, Petroleum Publishing Company, Tulsa, 1975.

Silbergh, M. and Brons, F.: "Profitability Analysis—Where Are We Now?" *Journal of Petroleum Technology*, January 1972, pp. 90–100.

Stermole, F. J.: *Economic Evaluations and Investment Decision Methods*, Third Edition, Investment Evaluations Corporation, Golden, Colorado, 1980.

7

VALUATION OF OIL AND GAS PROPERTIES

7.1 INTRODUCTION

The value of an oil and/or gas property may be defined by a number of parameters, but the most unequivocal definition of value is the sale price or fair market value. The fair market value is the price at which a property will change hands between a willing buyer and a willing seller, both of whom are knowledgeable in regard to the property and neither of whom is under any compulsion to buy or sell. Valuation or appraisal is defined as the establishment of a price.

There are a variety of valuation methods common in business. Valuation of producing properties offer particular difficulty because value is not necessarily proportional to investment, acreage, or even production rate. Three methods that the petroleum engineer, geologist, and manager may encounter are:

1. *Actual Cost or Replacement Value.* Actual cost or replacement value is often used to determine a price for auxiliary facilities, for example, the purchase of a compressor or of an unproductive well for waste water disposal. This method is not applicable to most producing properties because neither the acquisition cost nor the cost to develop the property reflects the value of the property at a later date. In recent years, many properties have increased significantly in value because of oil and gas price changes, improved recovery techniques, and increasing demand for oil and gas supplies. On the other hand, some properties have declined in value with declining production rates.

2. *Sale of Similar Properties.* Determination of price by analogy with similar properties, when properly done, gives the fair market value of the property. Transactions involving parties with special reasons to buy or sell must be eliminated from consideration, and great care must be exercised to ensure that only analogous properties are compared. Even though this method is not completely satisfactory, economists in many banks, consulting firms, and in

most major companies keep a close watch on actual transactions to determine industry trends.

3. *Capitalization of Income (Engineering Evaluation Methods).* In this method, future income from a property or project is forecast. The maximum price is calculated so that the earnings from the property will provide a satisfactory return on the investment (price) considering the risks. This is profitability analysis by the net present value profit method. The engineering evaluation method provides the most reliable guide to *value*, but judgment and experience remain the essential ingredients. It is used almost universally by major companies in the oil and gas industry. The rest of this chapter will be devoted to the mechanics of the engineering evaluation method and a discussion of other profitability criteria.

7.2 PRESENT VALUE OF DECLINING INCOME

The present value of future production is calculated by multiplying future production volumes with a discount factor. It is easy to relate production to money if the simplifying assumption is made that the net revenue per unit of production remains constant. The volumes obtained in this manner are called the present value production.

For an exponential production decline the production rate after t years is given by

$$q = q_i e^{-Dt} \tag{7.1}$$

where D is the constant decline rate per year. The present value of this production volume (q^*) is obtained by multiplying this amount with the discount factor for single end of year payments and continuous compounding $D_{sp}''' = e^{-jt}$. Thus,

$$q^* = q_i e^{-Dt} e^{-jt} = q_i e^{-(D+j)t} \tag{7.2}$$

This is equivalent to an exponentially declining production with an instantaneous decline rate of $(D + j)$.

The total present value production (Q_D^*) for the period $t = 0$ to $t = n$ is obtained by integrating Eq. 7.2:

$$Q_D^* = \int_0^n q_i e^{-(D+j)t} \, dt = \frac{q_i}{D + j} [1 - e^{-(D+j)n}] \tag{7.3}$$

The discount factor (D_e) for an exponentially declining income is defined as the number with which the undiscounted cumulative production has to be multiplied to arrive at the present value of this production, that is,

$$D_e Q_D = Q_D^* \tag{7.4}$$

From Eqs. 3.15 and 7.3,

$$D_e = \frac{D}{D+j} \frac{1 - e^{-(D+j)n}}{1 - e^{-Dn}} \tag{7.5}$$

which for large values of n simplies to

$$D_e = \frac{D}{D+j} \tag{7.6}$$

As explained in Chapter 3, often the assumption that the decline is exponential (constant percentage) throughout the life gives too conservative an estimate of the economic ultimate production. Under these conditions, other decline relationships like hyperbolic and exponential declines may be assumed.

7.3 PROFITABILITY OF A DEVELOPMENT WELL

In order to obtain authority for expenditure (A.F.E.) to drill an oil or gas well in a field being developed, an economic evaluation of the prospect must be made and the profit indicators (*NPV* or *DCFROR*) must be such that the venture is financially attractive enough. Other than good geological knowledge, there is very little information available, since the well has not been drilled and surrounding wells usually are still at an early stage of their producing life.

However, by studying the performance of surrounding wells, we can estimate the initial production rate of the well. Also empirical data in similar oil and gas fields may be helpful in estimating recoverable reserves. This information together with financial data can be used to determine the profitability of the new well.

Consider the case of one well in a lease block with good geological knowledge. The initial costs involved are lease bonus, seismic (geological and geophysical), rentals and taxes, and drilling costs. All initial costs are committed before the outcome is known, and are lost with a dry well. If the outcome is a producer, additional costs like well completion cost, lease royalty, state and local taxes, operating cost, lease rental, and federal income taxes will be incurred. There will be revenue from the sale of oil and gas. This should be enough to pay for the costs mentioned and guarantee a reasonable profit.

The main unknown is the time pattern of revenue. This may be estimated by making a few assumptions. If the lease area (A) and net pay thickness (h) are known from good geological information, and the recovery factor (E_R) is estimated by studying similar reservoirs, then the reserves are given for example as

$$Q(\text{bbl}) = A(\text{acres})h(\text{ft})E_R(\text{bbl/acre-ft}) \tag{7.7}$$

With a knowledge of the decline rate per year (D), the initial unrestricted flow rate can be estimated as

$$q_i = DQ \tag{7.8}$$

For good reservoir management the well should not be allowed to flow unrestricted. Suppose the maximum efficient rate *(MER)* or well allowable is q_c. The production pattern of the well will be as shown in Fig. 7.1. The well flows at an allowable rate for t_c years and then declines exponentially for t_d years before reaching the economic production rate, q_a.

The value of t_c may be obtained by an independent estimate or by equating the constant rate production to the production by exponential decline if the rate were unrestricted:

$$t_c = \frac{q_i - q_c}{Dq_c} \tag{7.9}$$

The time on decline (t_d) may be calculated with a knowledge of the economic limit.

Present Value Calculation

The present value factor depends on the time pattern of money and the discount rate. For continuous compounding and continuous and constant flow of money, Eq. 5.45 gives the present value factor. Thus the present value of operating cost is

$$PVOC = C_o(PVF_{oc}) \tag{7.10}$$

or

$$PVOC = \frac{C_o(1 - e^{-jt_a})}{j} \tag{7.11}$$

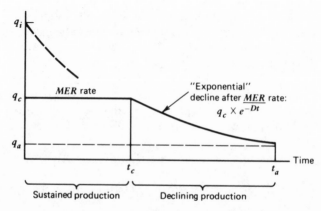

Figure 7.1 Production pattern of development well.

where C_o is the operating cost per year. Similarly the present value of constant rate production is

$$PVR_c = \frac{q_c(1 - e^{-jt_c})}{j} \qquad (7.12)$$

where q_c is the production rate in barrels per year. This can easily be converted to money values by multiplying by the net revenue per barrel.

The present value of exponentially declining production may be obtained from Eq. 7.3, shifted to time zero:

$$PVR_d = \frac{q_c}{D + j} \, e^{-jt_c}[1 - e^{-(D+j)t_d}] \qquad (7.13)$$

The present value of the composite production may be obtained by summing Eqs. 7.12 and 7.13:

$$PVR = \frac{q_c}{j}\left[(1 - e^{-jt_c}) + \left(\frac{j}{D + j}\right) e^{-jt_c}(1 - e^{-(D+j)t_d}) \right] \qquad (7.14)$$

It is possible to obtain a composite present value factor *(PVF)* for this production pattern such that

$$PVR = (\text{reserves})(PVF) \qquad (7.15)$$

The recoverable reserves are given by

$$Q = q_c t_c + \frac{q_c}{D} (1 - e^{-Dt_d}) \qquad (7.16)$$

$$= \frac{q_i - q_c}{D} + \frac{q_c}{D} (1 - e^{-Dt_d})$$

$$= \frac{q_i}{D}\left(1 - \frac{q_c}{q_i} e^{-Dt_d}\right)$$

or

$$Q = \frac{q_i}{D}\left(1 - \frac{q_a}{q_i}\right) \qquad (7.17)$$

Since $q_a/q_i \ll 1$,

$$Q \simeq \frac{q_i}{D} \qquad (7.18)$$

From Eqs. 7.14 and 7.18 the formula for the composite present value based on continuous compounding and recovery is

$$PVF = \left(\frac{q_c}{q_i}\right)\left(\frac{D}{j}\right)\left[(1 - e^{-jt_c}) + \left(\frac{j}{D+j}\right) e^{-jt_c}(1 - e^{-(D+j)t_d})\right] \quad (7.19)$$

If there is no period of prorated production, $(q_c/q_i) = 1$ and $t_c = 0$, and Eq. 7.19 simplifies to,

$$PVF = \left(\frac{D}{D+j}\right)[1 - e^{-(D+j)t_d}] \quad (7.20)$$

Example 7.1.   Consider the case of one well in an oil lease block. The reservoir data are as follows:

Lease area, A = 160 acres

Net pay thickness, h = 100 feet

Recovery, E_R = 80 bbl/acre-ft

Decline rate, D = 25%/year

MER rate, q_c = 500 bbl/day

Other pertinent information are:

Working interest = 100%

Royalty = 12½%

State and local taxes = 7%

Selling price of crude oil= $30/bbl

Operating cost = $5000/month

Drilling and Completion Costs = $2,000,000

Calculate on a before tax basis:

(a) Net present value of the oil to be produced at discovery time.

(b) Net present value at decision time if there is a six month lag from decision time to discovery time.

Solution
The reserves are calculated from Eq. 7.7:

$$Q = (160 \text{ acres})(100 \text{ ft})(80 \text{ bbl/acre-ft})$$

$$= 1,280,000 \text{ bbl}$$

The initial (unrestricted) flow rate is (using Eq. 7.8):

$$q_i = (0.25)(1,280,000)/(365)$$

$$= 877 \text{ bbl/day}$$

However, the *MER* rate, $q_c = 500$ bbl/day. The production pattern is as shown in Fig. 7.1.

The time on allowable production is given by Eq. 7.9:

$$t_c = \frac{877 - 500}{(0.25)(500)} = 3.01 \text{ years}$$

The net revenue per barrel is

$$V = \left(1 - \frac{1}{8}\right)(1 - 0.07)(\$30) = \$24.41/\text{bbl}$$

The economic limit is given by

$$q_a V = \text{operating cost}$$

or

$$q_a = \frac{\$5000/\text{month}}{(\$24.41/\text{bbl})(30.4 \text{ days/month})}$$

$$q_a = 6.74 \text{ bbl/day}$$

The abandonment time (using Eq. 3.29) is

$$t_a = t_c + t_d = t_c + \frac{1}{D} \ln\left(\frac{q_c}{q_a}\right)$$

$$= 3.01 + \frac{1}{0.25} \ln\left(\frac{500}{6.74}\right) = 20.24 \text{ years}$$

The present value factor using a discount rate of 15 percent ($j = 0.14$) is calculated from Eq. 7.19:

$$PVF = \left(\frac{500}{877}\right)\left(\frac{0.25}{0.14}\right)[(1 - e^{-0.4214})$$

$$+ \left(\frac{0.14}{0.39}\right) e^{-0.4214}(1 - e^{-6.7197})]$$

$$= 0.5896$$

The present value of the revenue is

$$PVR = Q(V)(PFV)$$
$$= (1{,}280{,}000)(\$24.41)(0.5896)$$
$$= \$18{,}421{,}934$$

The present value of operating costs (using Eq. 7.11) is

$$PVOC = \frac{(\$5000)(12)}{0.14}(1 - e^{-2.8336})$$
$$= \$403{,}371$$

Drilling and completion costs $= \$2{,}000{,}000$

(a) The net present value at discovery time is

$$NPV = \$(18{,}421{,}934 - 403{,}371 - 2{,}000{,}000)$$
$$= \$15{,}018{,}563$$

(b) If the decision time lags the discovery time by six months, the net present value at decision time is

$$NPV_{DT} = \$16{,}018{,}563[e^{-(0.14)(0.5)}]$$
$$= \$14{,}935{,}609$$

Example 7.2.   A new well is to be drilled in a developed area. The production data for a typical well in the area is shown in Fig. 7.2. These are used to forecast the production of the new well over an 11 year life. The initial production rate for wells in the area is below the allowable rate and, thus, there is no period of constant rate production. Other data are as follows:

Crude oil price $= \$30/$bbl

Royalty $= 20\%$

Operating cost $= \$1500/$well/month

Mineral tax $= 7\%$ of gross working interest income

Investment $= \$440{,}000$

Determine if this well is a financially attractive investment on a before tax basis.

Solution
The economic evaluation on a before tax basis is shown in Table 7.1. The *DCFROR* is slightly higher than 50 percent. A quick estimate of other profit indicators is presented below.

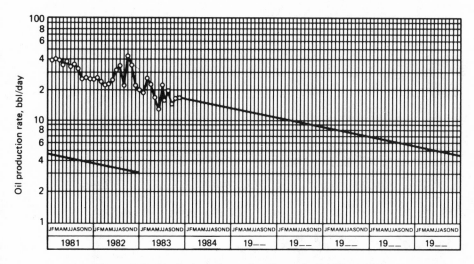

Figure 7.2   Typical producer for Example Problem 7.2.

Example 7.3.   An offshore well is to be drilled in 50 feet of water. The well will produce 660,300 barrels of crude oil at an average price of $30.64 per barrel, 16,500 barrels of natural gas liquids at an average price of $30.56 per barrel, and 443,500 Mscf of residue or dry gas at an average price of $1.64 per Mscf. The production forecast is given in Table 7.2. Well costs include the following:

Rig rentals, services, mud, etc.	$1,190,800
Casing, tubing, packers, christmas tree, etc.	699,200
Drilling overhead	200,000
Contingency costs (deepening, treating, and fishing)	210,000
Total	$2,300,000

It is estimated that there will be an additional expenditure of $1,040,000 in the sixth year for remedial work and redrilling. Facilities investments include the following:

Caisson and navigation aids	$ 300,000
Helicopter deck	140,000
High pressure flow line for gas lift and well shut-in devices	316,800
Prorate for tanks, separators, and other processing equipment	443,200
Total facilities cost	$1,200,000

Other data are:

Corporate tax rate	51%
Basic royalty	16.67%
Overriding royalty	3.33%

Evaluate the attractiveness of this investment.

TABLE 7.1
Estimated Future Net Revenue

Year	Annual Production (bbl)	Gross Income ($)	Royalty ($)	Gross Working Interest Income ($)	Operating Costs ($)	Mineral Tax $	Net Operating Income ($)	Midyear Discount Factor @ 50%	Present Value @ 50%
1	12,400	372,000	74,400	297,600	18,000	20,700	258,000	0.8165	211,400
2	11,300	339,000	67,800	271,200	18,000	18,900	234,300	0.5443	127,500
3	8,000	240,000	48,000	192,000	18,000	13,500	160,500	0.3629	58,200
4	5,300	159,000	31,800	127,200	18,000	8,700	100,500	0.2419	24,300
5	4,200	126,000	25,200	100,800	18,000	7,200	75,600	0.1613	12,200
6	3,300	99,000	19,800	79,200	18,000	5,700	55,500	0.1075	6,000
7	2,600	78,000	15,600	62,400	18,000	4,500	39,900	0.0717	2,900
8	2,100	63,000	12,600	50,400	18,000	3,600	28,800	0.0478	1,400
9	1,600	48,000	9,600	38,400	18,000	2,700	17,700	0.0319	600
10	1,300	39,000	7,800	31,200	18,000	2,100	11,100	0.0210	200
11	1,000	30,000	6,000	24,000	18,000	1,800	4,200	0.0142	—
	53,100	1,593,000	318,600	1,274,400	198,000	89,400	987,000		444,700

Investment = $440,000

ULTIMATE WELL INCOME—100% Working Interest

Gross income	$1,500,000 [1]
Royalty	(300,000) [2]
Gross working interest income	$1,200,000
Operating cost	(180,000) [3]
Mineral tax	(84,000) [4]
Net operating income	$ 936,000
Investment	$ 440,000
Profit/investment	2.1

Notes:
(1) 50,000 bbl. × $30.00/bbl
(2) 20% of gross income
(3) $18,000 per year × 10 years
(4) 7% of gross working interest income

Profit to investment ratio should be greater than 1.0 on a before tax basis.

MONTHLY WELL INCOME

Gross income	$ 27,000 [1]
Royalty	(5,400) [2]
Gross working interest income	$ 21,600
Operating cost	(1,500)
Mineral tax	(1,500) [3]
Net operating income	$ 18,600
Investment	$ 440,000
Payout	24 months

Notes:
(1) 900 bbl/month × $30.00/bbl
(2) 20% of gross income
(3) 7% of gross working interest income

The payout time should preferably not exceed 30 months.

TABLE 7.2
Production and Income Data

Year	100 Percent Production			Working Interest Salable Production			Working Interest Value			
	Crude Oil (bbl)	Natural Gas Liquids (bbl)	Residue or Dry Gas (mcf)	Crude Oil (bbl)	Natural Gas Liquids (bbls)	Residue or Dry Gas (mcf)	Crude Oil ($)	Natural Gas Liquids ($)	Residue or Dry Gas ($)	Total ($)
1984	105,850	2,646	71,094	84,680	2,117	56,875	2,594,596	64,728	92,820	2,752,144
1985	105,850	2,646	71,094	84,680	2,117	56,875	2,594,596	64,728	92,820	2,752,144
1986	105,850	2,646	71,094	84,680	2,117	56,875	2,594,596	64,728	92,820	2,752,144
1987	105,850	2,646	71,094	84,680	2,117	56,875	2,594,596	64,728	92,820	2,752,144
1988	79,205	1,980	53,198	63,364	1,584	42,558	1,941,472	48,436	69,456	2,059,369
1989	54,385	1,360	36,528	43,508	1,088	29,222	1,333,084	33,256	47,692	1,414,032
1990	44,530	1,113	29,909	35,624	891	23,927	1,091,520	27,232	39,048	1,157,800
1991	33,580	840	22,554	26,864	672	18,043	823,112	20,536	29,448	873,096
1992	25,185	630	16,915	20,148	504	13,532	617,336	15,400	22,084	654,820
Total	660,300	16,500	443,500	528,200	13,200	354,800	16,184,800	403,600	579,200	17,167,600

Solution

Table 7.2 gives the production and income data. Details of operating expense calculations are given in Table 7.3. Complete cash flow details are presented in Table 7.4. Details of *DCFROR* analysis are given in Table 7.5.

The profit indicators are summarized below:

$$
\begin{aligned}
\text{Initial investment} &= \$3,500,000 \\
\text{Project life} &= 9 \text{ years} \\
\text{Undiscounted profit} &= \$3,079,200 \\
\text{Profit/\$ invested} &= 0.88 \\
\text{Payout} &= 2.253 \text{ years} \\
DCFROR &= 38.4\%
\end{aligned}
$$

Net present value profit

$$
\begin{aligned}
\text{At } 12\% &= \$1,588,400 \\
\text{At } 15\% &= \$1,324,000 \\
\text{At } 20\% &= \$946,800
\end{aligned}
$$

7.4 SHORT CUT PROFITABILITY METHODS

Because of the complexity of the equations presented in Section 7.2, it is often advisable to use approximations that have been shown to give acceptable results.

TABLE 7.3
Operating Expense Calculation

	Operating Costs Including Department Overhead[a]				
Year Pre-	Producing Operating Costs ($)	Abandonment Costs ($)	Total Costs ($)	Mineral and Property Taxes ($)	Operating Income After Mineral and Property Taxes ($)
1984	613,200		613,200	197,516	1,941,428
1985	613,200		613,200	197,516	1,941,428
1986	613,200		613,200	197,516	1,941,428
1987	613,200		613,200	197,516	1,941,428
1988	613,200		613,200	147,796	1,298,368
1989	613,200		613,200	101,484	699,348
1990	613,200		613,200	83,092	461,504
1991	613,200		613,200	62,660	197,232
1992	613,200		613,200	46,996	−5,376
Total	5,518,800		5,518,800	1,232,000	10,416,800

[a]Department overhead is 5%.

TABLE 7.4
Income Tax Calculation

Year Pre-	Total WI Income (Including Salvage) ($)	Operating Expense[a] ($)	Depreciation ($)	Depletion Used[b] ($)	Intangible Investment ($)	Net Income for Income Tax Calculation ($)	Income Tax ($)	Net Operating Income After Tax ($)	Total Investment ($)	Net Income After All Investments
Pre-					1,600,800				3,500,000	−3,500,000
1984	2,752,144	810,716	345,656			−5,028	−192,484	2,133,912		2,133,912
1985	2,752,144	810,716	282,744			1,658,684	845,928	1,095,500		1,095,500
1986	2,752,144	810,716	231,284			1,710,140	872,172	1,069,256		1,069,256
1987	2,752,144	810,716	189,192			1,752,236	893,640	1,047,788		1,047,788
1988	2,059,364	766,996	154,760			1,143,608	583,240	715,128		715,128
1989	1,414,032	714,684	139,368		899,600	−339,620	−183,032	882,384	1,040,000	−157,616
1990	1,157,800	696,292	126,780			334,724	170,708	290,796		290,796
1991	873,096	675,860	103,708			93,528	47,700	149,536		149,536
1992	654,820	660,196	466,108			−471,484	−240,456	235,080		235,080
Total	17,167,600	6,750,800	2,039,600		2,500,400	5,876,800	2,797,600	7,619,200	4,540,000	3,079,200

[a]Includes operating cost, department overhead, mineral and property tax.
[b]No depletion allowance taken.

TABLE 7.5
Rate of Return Calculation

Year	Net Income After All Investments	Discount Factor[a] $i = 38\%$	Present Value Net Income at 38%	Discount Factor[a] $i = 39\%$	Present Value Net Income at 39%
Pre-	−3,500,000	1.0000	−3,500,000	1.0000	−3,500,000
1984	2,133,912	0.8513	1,816,600	0.8482	1,809,960
1985	1,095,500	0.6169	675,812	0.6102	668,484
1986	1,069,256	0.4470	477,956	0.4390	469,400
1987	1,047,788	0.3239	339,388	0.3158	330,920
1988	715,128	0.2347	167,852	0.2272	162,488
1989	−157,616	0.1701	−26,808	0.1635	−25,764
1990	290,796	0.1232	35,840	0.1176	34,196
1991	149,536	0.0893	13,356	0.0846	12,652
1992	235,080	0.0647	15,216	0.0609	14,308
Total	3,079,200		15,212		−23,356

Therefore, the rate of return ≃ 38.39% by interpolation

[a]Midyear.

Consider the time pattern of recovery shown in Fig. 7.1. During the period of constant or allowable production if payments are received at midyear, the present value of the allowable production is (using Eq. 5.25),

$$Q_c^* = q_c(v) \left[\frac{(1 + i)^{t_c} - 1}{i(1 + i)^{t_c - 0.5}} \right] \qquad (7.21)$$

where

Q_c^* = present value of the allowable or constant rate production, $

q_c = allowable production rate, bbl/year

v = net income per barrel, $/bbl

i = discount rate, %/year

t_c = time of allowable production, years

Similarly, during the period of exponential decline, the present value of declining production, discounted to the start of decline is

$$Q_D^* = Q_1 \left[\frac{(1 + i)^{0.5} - r^{t_d}/(1 + i)^{t_d - 0.5}}{1 + i - r} \right] \qquad (7.22)$$

where

Q_d^* = present value of the declining production at the start of the decline period, $

Q_1 = income during the first year of decline, $

r = ratio of production rates of successive years (Eq. 3.23)

t_d = number of years of production on exponential decline

If Q_D is the amount of recoverable reserves during the declining production, Eqs. 3.15 and 3.23 may be used to obtain the following relationship for the income during the first year of decline:

$$Q_1 = Q_D(v) \left[\frac{1 - r}{1 - r^{t_d}} \right] \tag{7.23}$$

The composite present value production at discovery time is given by

$$Q^* = Q_c^* + \frac{Q_D^*}{(1 + i)^{t_c}} \tag{7.24}$$

Example 7.4. A lease producing at an allowable rate yields a net income of $300,000 per year. This lease has a 10-year period of exponential decline following its five years of allowable production. The lease will be abandoned when the net income falls to $30,000 per year. Calculate the net present value of the production from this lease at discovery time if the discount rate is 15 percent per year.

Solution

$$q_c = \$300,000/\text{year}$$

$$q_a = \$30,000/\text{year}$$

$$t_d = 10 \text{ years}$$

From Eq. 3.23,

$$r^{t_d} = \frac{q_a}{q_c} = \frac{30,000}{300,000} = 0.10$$

$$r = (0.10)^{1/10} = 0.794$$

$$D = \frac{1}{t_d} \ln \left(\frac{q_c}{q_a} \right)$$

$$= \frac{1}{10} \ln \left(\frac{300,000}{30,000} \right) = 0.23 = 23\% \text{ per year}$$

The total undiscounted income during decline is

$$Q_D = \frac{q_c - q_a}{D} = \frac{\$300,000 - \$30,000}{0.23} = \$1,173,900$$

Income during the first year of decline is calculated from Eq. 7.23:

$$Q_1 = 1,173,900 \left(\frac{1 - 0.794}{1 - 0.10} \right) = \$269,700$$

Net present value of allowable production is given by Eq. 7.21:

$$Q_c^* = \$300,000 \left[\frac{(1.15)^5 - 1}{0.15(1.15)^{4.5}} \right] = \$1,078,400$$

The net present value of declining, discounted to the start of the decline period is given by Eq. 7.22:

$$Q_D^* = \$269,700 \left[\frac{(1.15)^{0.5} - 0.10/(1.15)^{9.5}}{1 + 0.15 - 0.794} \right] = \$792,300$$

The composite net present value production at discovery time is

$$Q^* = \$1,078,400 + \frac{\$792,300}{(1.15)^5} = \$1,472,300$$

Some graphical techniques exist for quick and approximate calculation of profit indicators. Figures 7.3 and 7.4 relate discount factor D_d, the initial to final production rate ratio q_i/q_a, and nj for exponential decline and hyperbolic decline (with $b = 0.5$).

Example 7.5.   An oil well that requires a total initial investment of $800,000 will generate a total undiscounted operating cash income of $1,600,000 in 10 years. If the ratio of initial to final production rates is 100, what is the *DCFROR*?

Solution
Discount factor

$$D_d = \frac{\$800,000}{\$1,600,000} = 0.50$$

From the exponential decline graph (Fig. 7.3), for $D_d = 0.50$ and $q_i/q_a = 100$, $nj = 4.7$. Since $n = 10$ years, $j = 4.7/10 = 0.47$, and the corresponding rate of return $i = e^j - 1 = 0.60$ or 60%.

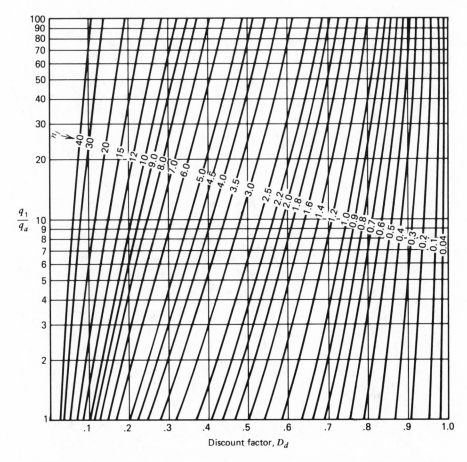

Figure 7.3 Discount factors for exponentially declining incomes (after Brons).

Example 7.6. An oil field producing unrestricted is for sale. The following information is available:

Present production rate	20,000 bbl/day
Production rate one year ago	24,000 bbl/day
Price of oil plus associated gas	$30/bbl
Royalty	12½%
Mineral tax	$2.50/bbl
Direct operating costs	$7000/day

Find the total undiscounted net operating income and the discounted net operating incomes using an opportunity rate of 15 percent. Consider both an exponentially declining production and a hyperbolically declining production with $b = 0.5$.

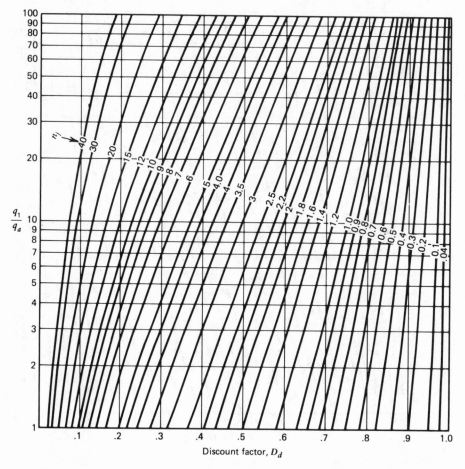

Figure 7.4   Discount factors for hyperbolically declining incomes ($b = 0.5$) (after Brons).

Solution
(a) Exponential decline

$$\frac{q_2}{q_1} = \frac{20,000}{24,000} = e^{-D}$$

Therefore,

$$D_d = 0.18 \quad \text{or} \quad 18\%/\text{year}$$

Gross working interest income = \$30 (0.875) = \$26.25/bbl
Net revenue = \$26.25 − \$2.50 = \$23.75/bbl

Economic limit of production is

$$q_a = \frac{\$7,000/\text{day}}{\$23.75/\text{bbl}} = 295 \text{ bbl/day}$$

$$\frac{q_i}{q_a} = \frac{20,000}{295} = 68$$

The remaining life (Eq. 3.29) is

$$t_a = \frac{1}{D} \ln \left(\frac{q_i}{q_a}\right) = 23 \text{ years}$$

The remaining economic reserves from Eq. 3.20 are:

$$Q_{Da} = \frac{q_i - q_a}{D} = \frac{365(20,000 - 295)}{0.18} = 40 \text{ million bbl}$$

Net revenue (undiscounted) = $23.75 (40,000,000) = $949.0 million
Total operating costs (undiscounted) = (365)(23)($7,000) = $58.8 million
Net operating income (undiscounted) = $(949 − 58.8) million = $890.2 million
Discount rate $i = 15\%$
Therefore,

$$j = 0.14 \quad \text{and} \quad nj = 3.21$$

From Fig. 7.3,

$$D_d(nj = 3.21, \frac{q_i}{q_a} = 68) = 0.58$$

$$D_c(nj = 3.21, \frac{q_i}{q_a} = 1) = 0.30$$

Discount net operating income

$$= \$(949)(0.58) - \$(58.8)(0.30) \text{ million}$$

$$= \$532.8 \text{ million}$$

(b) Hyperbolic decline ($b = 0.5$)
  From Eq. 3.75

$$\left(1 + \frac{D_i}{2} t\right)^2 = \frac{q_1}{q_2} = \frac{24,000}{20,000}$$

or

$$D_i = 0.19 = 19\%/\text{year}$$

Since $q_a = 295$ bbl/day and $q_i/q_a = 68$, using Eq. 3.77,

$$t_a = \frac{2(\sqrt{68} - 1)}{0.19} = 76 \text{ years}$$

The remaining economic reserves (Eq. 3.76) are

$$Q_{Da} = \frac{365(20,000 - \sqrt{(20,000)(295)})}{0.19} = 67.5 \text{ million bbl}$$

Net revenue (undiscounted) = \$23.75 (67.5 million) = \$1603 million
Total operating costs (undiscounted) = (365)(76)(\$7000) = \$194.2 million
Net operating income (undiscounted) = \$(1603 − 194.2) million = \$1408.8 million
For $j = 0.14$, $nj = 10.64$
From Fig. 7.4, $D_d(nj = 10.64, q_i/q_a = 68) = 0.38$
$\qquad\qquad D_c(nj = 10.64, q_i/q_a = 1) = 0.095$
The discounted net operating income

$$= \$(1603)(0.38) - \$(194.2)(0.095) = \$590.7 \text{ million}$$

Brons and Silbergh have presented nomographs that relate the total percent profit, discounted cash flow rate of return or internal rate of return, and payout time for three different income time patterns: constant rate, exponentially declining, and hyperbolically declining ($b = 1/2$). The effect of delay in the start of the income stream is considered also. These nomographs are shown in Figs. 7.5 to 7.8. These figures may be used as a general approximation and should not replace more accurate calculations. Each of these figures contain the following variables:

C = investment

I = total income from investment

$P = (I - C)/C$, total percent profit (undiscounted)

t_p = payout time, months or years

t_D = delay time between investment and start of income

j = nominal continuous rate of return

i = effective internal rate of return

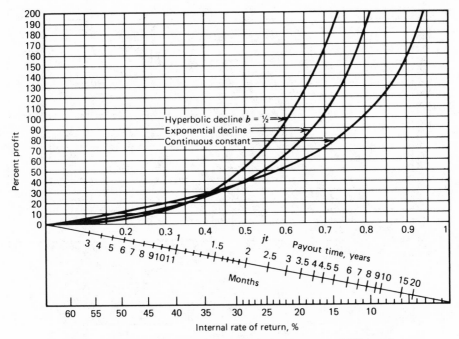

Figure 7.5 Profitability nomograph (After Brons and Silbergh).

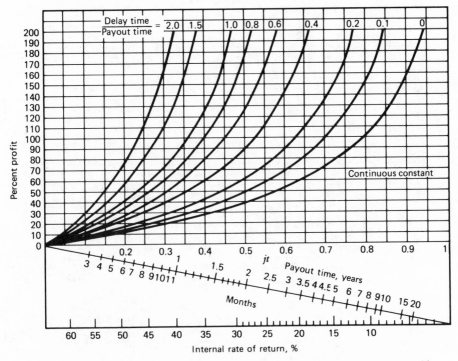

Figure 7.6 Profitability nomograph, constant rate (After Brons and Silbergh).

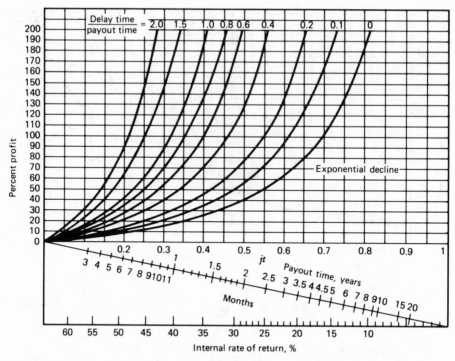

Figure 7.7 Profitability nomograph, exponential decline (After Brons and Silbergh).

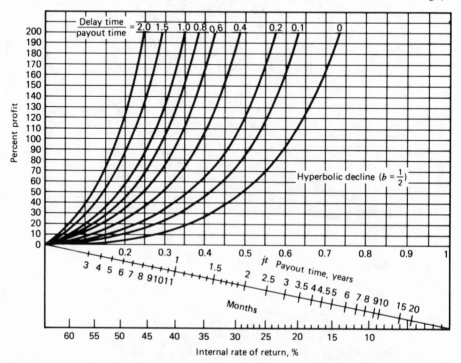

Figure 7.8 Profitability nomograph, hyperbolic decline (After Brons and Silbergh).

Example 7.7.   Consider investment in an oil-producing property that will be paid 24 months.

$$P = \frac{I - C}{C} = \frac{936,000 - 440,000}{440,000} = 113\%$$

Assuming exponential decline:

(*a*) What will the *DCFROR* be?

(*b*) What will be the *DCFROR* if the start of income is delayed by two years?

Solution

(*a*) From Fig. 7.7 and the curve for $t_D/t_P = 0$,

$$P = 113\%, \qquad t_P = 2 \text{ years}, \qquad \text{and} \qquad DCFROR \approx 43\%$$

(*b*) From Fig. 7.7 and the curve for $t_D/t_P = 2.0/2.0 = 1.0$,

$$P = 113\%, \qquad t_P = 2 \text{ years}, \qquad \text{and} \qquad DCFROR \approx 18\%$$

7.5   ACCELERATION PROJECTS

An acceleration project occurs when we speed up production that would otherwise be recovered at a slower rate or a later date. In a simplified form acceleration may be represented by a single investment C, resulting in an operating net income I at a relatively early date and losing the same operating net income I at a later date.

The total cash flow consists therefore of three distinct successive parts: negative (investment), positive, and again negative (loss). The last period of negative cash flow is unavoidable in a pure acceleration case. Obviously, there never would be a reason for such an acceleration project if it were considered on an undiscounted basis. The undiscounted profit and profit-to-investment ratio are both negative, making the project difficult to justify. We are basically sacrificing ultimate profit to get current income.

Acceleration projects are desirable from the standpoint of maximizing present value. However, problems may arise if the investment is significant and no *new* reserves are generated. Also, there is more than one *DCF* rate of return, which further complicates the interpretation of the profit indicators. The characteristic features of the cash flow and profit indicators of acceleration projects may be demonstrated by the following example.

Example 7.8.   Consider an existing project with an annual cash generation of $8000 expected to last 15 years. If an additional single investment of $10,000 is made, the cash flow will be accelerated to $12,000 per year but would last only 10 years, since no new cash was generated. Analyze the profitability of this acceleration project.

Solution

The cash flow table of the existing project A, acceleration project B, and the incremental $(B - A)$ are shown in Table 7.6. The incremental cash flow shows an increase in cash generation of $4000 per year during the first 10 years and an annual decrease in cash generation of $8000 during the final five years. The acceleration project shows a low of $10,000 and a $P/\$$ invested of -1.0! Is this an attractive investment?

A graphical representation of the cash flow is shown in Fig. 7.9. The cumulative undiscounted cash flow starts from $-\$10,000$, reaches a maximum of $30,000 at the tenth year and then decreases to $-\$10,000$ at the end of the project life. The discounted cumulative cash flow curves are also shown in Fig. 7.9. At an end of year discount rate of 5 percent the ultimate present value cash flow becomes zero. According to the definition of the *DCF* rate of return, the *DCF* rate of return for this project is 5 percent.

At discount rates higher than 5 percent, the cumulative present value becomes positive, reaches a maximum value at 15 percent, and decreases to zero again at 45 percent. Thus 45 percent could also be defined as the *DCF* rate of return for the project. The present value profile for this project is shown in Fig. 7.10. The curve crosses the discount rate axis twice, giving rise to two *DCF* rates of return. Although this fails to provide a unique answer, the present value profile could be used effectively to determine the profitability of this acceleration proj-

TABLE 7.6
Acceleration Project

Year	A Existing Anticipated Income	B Proposed Accelerated Income	B − A Incremental Undiscounted
0	—	10,000	10,000
1	8,000	12,000	4,000
2	8,000	12,000	4,000
3	8,000	12,000	4,000
4	8,000	12,000	4,000
5	8,000	12,000	4,000
6	8,000	12,000	4,000
7	8,000	12,000	4,000
8	8,000	12,000	4,000
9	8,000	12,000	4,000
10	8,000	12,000	4,000
11	8,000	—	8,000
12	8,000	—	8,000
13	8,000	—	8,000
14	8,000	—	8,000
15	8,000	—	8,000
Total	120,000	110,000	10,000

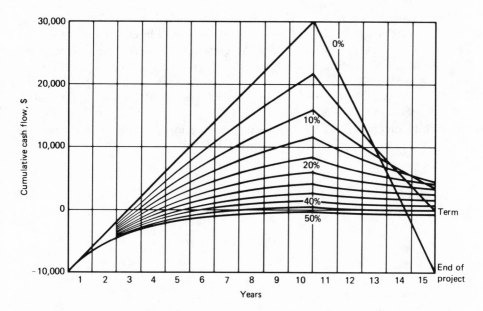

Figure 7.9   Acceleration project showing cumulative cash flow at different discount rates.

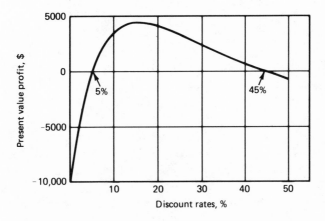

Figure 7.10   Present value profile for acceleration project.

ect. Figure 7.10 indicates that the project is acceptable at opportunity rates between 5 and 45 percent.

REFERENCES

Brons, F.: "Profitability Analysis for Petroleum Engineers," Lecture Notes for SPE Short Course, Houston, 1979.

Brons, F. and Silbergh, M.: "The Relation of Earning Power to Other Profitability Criteria," *Journal of Petroleum Technology*, March 1964, p. 269.

Henry, A. J.:"Appraisal of Income Acceleration Projects," paper SPE 3460, Annual Fall Meeting of the Society of Petroleum Engineers, New Orleans, October, 1971.

McCray, A. W.: *Petroleum Evaluations and Economic Decisions*, Prentice-Hall, Englewood Cliffs, New Jersey, 1975.

Stanford University: "Lecture Notes on Engineering Valuation and Appraisal of Oil and Gas Properties," Stanford, California, 1976.

8

ANALYSIS OF RISK AND UNCERTAINTY

8.1 INTRODUCTION

So far in our analysis of profitability criteria, risk and uncertainty in the input data have been left out of consideration. This does not mean that risk and uncertainty were ignored, but rather that in such cases the values of the parameters used were assumed to be the *best estimate* of these values. However, the spread of values around these best estimates may influence our decisions in arriving at a choice between ventures.

Consider that an opportunity exists for acquiring a block of open acreage or an exploration concession for a considerable sum of money. An analysis must first be made to ascertain if there is a favorable chance for a discovery that will offset all necessary financial outlays and yield an adequate return on expended capital.

The traditional approach in performing this analysis (Fig. 8.1) is to first estimate the possible producible reserves for the prospective acreage. This involves making a "best estimate" of values for the individual petrophysical parameters that appear in the familiar volumetric reserve formula. This "best estimate" or "most likely" values are often used in calculating "base case" evaluations. Through a combination of experience, intuition, judgment, and consensus the explorationist and manager must then decide if the profitability is large enough.

This type of one-point analysis gives acceptable results in many cases where the uncertainty is not large and the profit margin is not critical. Unfortunately, such projects are becoming scarcer in many areas of the oil and gas industry.

The main limitations of the traditional one-point analysis are:

1. It does not allow for quantitative assessment of risk and uncertainty.

2. It is an "either/or" analysis (discovery or a dry well). It cannot account for other possible reserve levels, economic uncertainties, and the like.

191

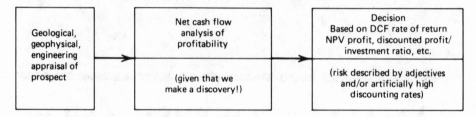

Figure 8.1 Traditional approach for analyzing drilling prospects. (After Newendorp, 1975).

3. It is difficult to characterize risk with qualitative assessments such as the following statements:

    (a) We have a good shot at the formation.
    (b) Control is fair.
    (c) This is about as sure as a wildcat can be.
    (d) This is pretty risky—moderately risky—high risk, and so on.

Sensitivity Analysis

Assuming that the wildcat is successful, a preliminary determination must be made to see whether or not the "most likely" reserves justify a development effect. Sensitivities of the outcomes to variations in the "base case" data are generated by evaluating "low estimate" and "high estimate" cases. From these different cash flows and various financial performances, indicators are computed and checked against minimum requirements for entering into new ventures. Sensitivity analyses to show the effect of changes in the key assumptions is the first step in risk analysis beyond a one-point prediction. They help answer the question, "What if?"

Sensitivity analysis explores the relative effects on the economic viability of the project as a whole on changes in items contributing to cash flows. It pinpoints areas that are most critical in terms of any uncertainty and indicates where confidence in estimates is most vital. Sensitivity analysis is used to explore the effects on a project of uncertainty in different areas, but does not attempt to quantify the relative uncertainty in different areas. This is still left to subjective qualitative interpretation of results. The next step in trying to deal systematically and rationally with uncertainty in economic evaluation is to attempt to quantify the uncertainty in terms of subjective probabilities of possible alternative outcomes.

Subjective Probabilities

To aid in decision making under conditions of uncertainty, we assign a chance or probability of occurrence to the possibility of a certain outcome. These subjective probabilities are commonly used to quantify uncertainty. How this is done, and how valid these probabilities are, depend on the circumstances. It is important to realize that they are merely a way of expressing personal feelings about uncertainty and have no objective reality. Objective probabilities can be demonstrated and tested. For example, the probability of getting a head can be found by tossing a coin a large number of times. There is no similar basis for the probabilities about the future of a project, since there are not enough similar projects to reach a satisfactory conclusion, and each project differs in some way from previous ones.

Sometimes subjective probabilities are based on observed past frequencies of occurrence of similar events, and on the Baysian hypothesis that these frequencies may occur in the future. Or they may be based on the subjective feeling of the person making the study, which may be based on past experience, study of existing circumstances, or just intuition. The advantage of using subjective probabilities as means of expressing opinions about uncertain situations is not only that the numbers have objective reality, but that they enable the consequences of these opinions to be explored logically and rationally.

Obtaining Probabilities

Quantitative statements about risk and uncertainty are given as numerical probabilities, or likelihoods of occurrence. Probabilities are decimal fractions in the interval from zero to 1.0. An event or outcome that is certain to occur has a probability of occurrence of 1.0. As the probability approaches zero, the event becomes increasingly less likely to occur. An event that cannot occur has a probability of occurrence of zero. The probability of a head on the flip of a fair coin is 0.5. Assigning probabilities of occurrence to the various outcomes of a petroleum venture requires the cooperative judgment and skills of geologists, engineers, and geophysicists.

The types of risks we need to quantify or assess may include:

1. Risk of an exploratory or development dry hole.

2. Political risk, economic risk.

3. Risk relating to future oil and gas prices.

4. Risk of storm damage to offshore installations.

5. Risk that a discovery will not be large enough to recover initial exploratory costs.

6. Risk of at least a given number of discoveries in a multiwell drilling program.

7. Environmental risk.

8. Risk of gambler's ruin.

Certain characteristics are unique to the petroleum industry. (1) We cannot explicitly describe the process that originally generated the probability distribution of petroleum accumulations; thus we are not able to develop an exact probabilistic model. (2) The drilling of a sequence of wells is a series of dependent events based on sampling *without* replacement, but models that have been proposed are usually based on the notion of independent events. (3) Probability estimates must often be made on the basis of very little or no statistical data or experience. Additional data in petroleum exploration are usually from additional wells or seismic, and we normally can't afford to delay decisions until there is sufficient amount of data upon which to base our probability estimates.

Even though the above characteristics make it difficult to make decisions in the petroleum industry, the following methods will help us to obtain probability estimates:

1. Subjective probability estimates are a common way of expressing the degree of risk and uncertainty. The estimates represent personal opinion based on available statistical data or on the analyst's feeling.

2. Risk is often treated by raising the minimum no-risk standards required to accept a project. By this approach management is attempting to protect against uncertainty by raising the investment "cut-off" rate. This does not consider the varying levels of risk between competing investments.

3. The use of past success ratios is the most common method of estimating probabilities. This implies independent events and sampling *with* replacement.

4. Models using ultimate reserve distributions are based on the hypothesis that the probability density function of reserves remaining to be discovered at any time is proportional to the area between the ultimate reserve distribution and the distribution of reserves discovered to date. It is assumed that the ultimate reserve distribution is approximately log normal.

5. The binomial, multinomial, and hypergeometric distributions are used in multiwell drilling programs. In using these probability distributions it must be decided whether the events are dependent or independent.

The numbers we get with any of these methods are not necessarily ends in themselves. There are no absolutes in risk analysis.

Weighted Average Sensitivity

Sensitivity analysis can be improved one step further by weighting the possible outcomes. Probability of occurrence is assigned to each outcome. By weighting the different outcomes by their probability of occurrence, a probability-weighted outcome known as the Expected Value *(EV)* is obtained. These three steps toward risk analysis are illustrated in Table 8.1.

This bring us to the risk analysis approach to economic evaluation (Fig. 8.2). It incorporates probabilistic geological, engineering, and economic data to generate frequency distributions of basic economic and physical outcomes of exploiting new reserves. It is based on the concept of maximizing monetary expectations, and thus uses risk-weighted expected values. The starting point of this approach is still the same geological–geophysical studies. There is no need for the geologist and geophysicist to feel threatened. However, an understanding of certain probability and statistical concepts associated with risk analysis and the expected value concept, by both the decision makers and the explorationists, is required.

TABLE 8.1
Three Steps Toward Risk Analysis

1. One-point analysis (best estimate)

Reserves	Net Profit
400 Mbbl	$6,000 M

2. Sensitivity analysis (what if?)

	Reserves	Net Profit
Dry hole	0	($ 500 M)
	200 Mbbl	3,000 M
Base case	400 Mbbl	6,000 M
	600 Mbbl	9,000 M

3. Weighted average sensitivity (expected value):

	Reserves	Net profit	×	Probability	=	EV
Dry hole	0	($ 500 M)		0.15		($ 75 M)
	200 Mbbl	3,000 M		0.30		900 M
Base case	400 Mbbl	6,000 M		0.35		2,100 M
	600 Mbbl	9,000 M		0.20		1,800 M
				EMV	=	$4,725 M

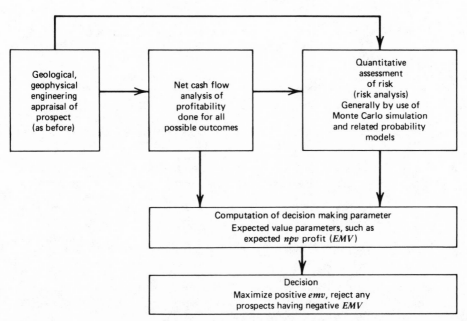

Figure 8.2 Proposed approach that accounts for risk and uncertainty. (After Newendorp, 1975).

Decision analysis methods provide new and much more comprehensive ways to evaluate and compare the degree of risk and uncertainty associated with each investment choice. The net result is that the decision maker is given a clearer insight of potential profitability and the likelihoods of achieving various levels of profitability than older, less formal methods of investment analysis. Because of such factors as rising drilling costs, the need to search for petroleum in deeper horizons or in remote areas of the world, and increasing government control, most petroleum exploration decision makers are no longer satisfied to base their decisions on experience, intuition, rules of thumb, or similar approaches. Instead, they recognize that better ways to evaluate and compare drilling investment strategies are needed. Decision analysis is fulfilling this need.

Decision analysis involves the following steps:

1. Define possible outcomes that could occur for each of the available decision choices, or alternatives.

2. Evaluate profit or loss (or any other measure of value or worth) for each outcome.

3. Determine or estimate the probability of occurrence of each possible outcome.

4. Compute a weighted average profit (or measure of value) for each decision choice, where the weighting factors are the respective probabilities of occurrence of each outcome. This weighted average profit is called the expected value of the decision alternative and is the comparative criterion used to accept or reject the alternative.

In considering the merits of decision analysis it is important to note several distinct advantages that the risk analysis approach has over the less formal procedures used in the past:

1. Decision analysis forces a more explicit look at the possible outcomes that could occur if the decision maker accepts a given prospect.

2. Certain techniques of decision analysis provide excellent ways to evaluate the sensitivity of various factors relating to overall worth.

3. Decision analysis provides a means to compare the relative desirability of drilling prospects having varying degrees of risk and uncertainty.

4. Decision analysis is a convenient and unambiguous way to communicate judgments about risk and uncertainty.

5. Exceedingly complex investment options can be analyzed using decision analysis techniques.

However, decision analysis will not eliminate risk in decision making. Its main utility is to give the decision maker a better understanding of risk and uncertainty. These methods are intended to supplement, rather than replace the necessary judgment of managers, geologists, and engineers in the petroleum industry.

8.2   MATHEMATICAL EXPECTATION

The cornerstone of decision analysis is the expected value concept—a method of combining profitability estimates with quantitative estimates of risk to yield a risk-adjusted decision criterion. Virtually all formal strategies for decision making given uncertainty rest on the expected value concept.

Commonly used "profit-indicators" such as rate of return, payout, and profit-to-investment ratios are usually expressed in dollars only, with no consideration for risk. But risk is an ever present (and often critical) factor in drilling investment decisions—and must be considered among our "profit-indicators" for effective decision making.

Risk and uncertainties in petroleum exploration result because:

1. Geologists and engineers are not able to measure or define specific values of factors contributing to overall profit or loss at the time of decision.

2. Uncertainties about future events that could affect the timing and/or size of the projected cash flows from the prospects.

Expected Monetary Value

The expected value of an outcome is the product obtained by multiplying the probability of occurrence of the outcome and the conditional value (or worth) that is received if the outcome occurs. The value received if an outcome occurs can be expressed in various ways: monetary profits and losses, opportunity losses, preference or utility values based on associated profits and losses, and so on. For the special and most common case where the values received are expressed as monetary profits or losses the product is usually called expected monetary value of the outcome, or *EMV*. For this special case monetary losses are usually expressed as negative profits.

The expected value of an outcome is the algebraic sum of the expected values of each possible outcome that could occur if the decision alternative is accepted. It can be positive, zero, or negative. It is the numerical criterion used to compare competing decision choices.

Example 8.1.   To illustrate these principles of expected monetary value, consider the gamble that consists of flipping a fair coin. The rewards are a win of $5 if the result is a head and a loss of $2 if the result is a tail. Should you accept the gamble?

Solution

Outcome	Probability of Occurrence	Decision Alternative: Accept the Gamble	
		Conditional Value Received	Expected Value of Outcome
Head	0.50	+$5.00	+$2.50
Tail	0.50	−$2.00	−$1.00
			+$1.50

The expected value of the decision alternative "accept the gamble" is $1.50 compared to $0 if we don't accept. We therefore accept the gamble.

The decision rule for expected monetary value choices is: when choosing among several mutually exclusive decision alternatives, select the alternative having the highest positive expected monetary value *(EMV)*. Mutually exclusive means that acceptance of one alternative precludes acceptance of any other alternative. The decision maker must select only one of the available alternatives. The interpretation of expected value of a decision alternative is the average monetary profit per decision that would be realized if he accepted the alternative over a series of repeated trials.

The properties and rules of mathematical expectation are:

1. Any number of outcomes can be considered so long as the probabilities for all outcomes listed sum up to 1.0.

2. Any number of decision alternatives can be considered.

3. The conditional values can be expressed as before tax or after tax monetary values, and can be undiscounted or discounted net present value *(NPV)* profits.

4. The decision rule for the special case of the *EMV* implies that the decision maker is totally impartial to money and the magnitude of money involved in the gamble. This assumption may not be satisfied in some instances.

Example 8.2. Let us apply the concept of *EMV* to investment decisions. A company is considering the purchase of 320 net acres in a proposed 640 acre oil unit. Three decision alternatives are available to the company:

1. Participate in the unit with nonoperating 50% working interest *(WI)*.

2. Farm out, but retain ⅛ of ⅞ overriding royalty interest *(ORI)*.

3. Be carried under penalty with a back-in privilege (50% *WI*) after recovery of 150% of investment by participating parties.

The possible outcomes, based on detailed geological and engineering analyses of the prospect and surrounding wells, are shown below. This is called the payoff table.

Possible Outcomes	Probability of Occurrence	NPV If Company Participates	NPV If Company Farms Out	NPV with Back In
Dry hole	0.30	−$ 40,000	0	0
200 Mbbl	0.25	+$ 50,000	+$ 12,000	+$ 12,000
400 Mbbl	0.25	+$300,000	+$ 60,000	+$145,000
800 Mbbl	0.10	+$700,000	+$120,000	+$400,000
1000 Mbbl	0.10	+$800,000	+$130,000	+$500,000
	1.00			

Assuming a minimum acceptable rate of return of 10 percent, which of the three decision alternatives should the company accept?

Solution

The expected value computation is presented in Table 8.2. The results are summarized below:

Decision Alternative	Expected Net Present Value Profit
Drill	$225,000—Highest *EV*
Farm out	$43,000
Back in option	$129,000

Using the strategy of maximizing expected monetary profits, the alternative "Drill with 50% *WI*" has the highest expected net present value profit and would be the preferred choice.

Meaning of Expected Value

Expected value or mathematical expectation means in the long run. The calculated *EMV* represents the average profit per decision that would be realized if the decision maker accepted the alternative over a series of repeated trials. It is not guaranteed or even suggested that the next outcome will equal the *EMV*. This constitutes a problem in oil and gas exploration. Each new well (or prospect) is a unique, one-time decision. Is the expected value strategy actually valid in the oil and gas industry?

The key point is that the decision at hand is presumably one of many decisions under uncertainty presented to the decision maker. If he consistently selects the alternative having highest positive expected monetary value, his total net gain from all decisions will be higher than his gain from any alternative strategy for selecting decisions under uncertainty. This is true even though each specific decision is a different drilling prospect, with different probabilities and conditional profitabilities. The *EMV* is not so much a measure of actual expected profit of each (unique) decision but, rather, a strategy for making decisions. The strategy must be consistently used, and many decisions must be made on the basis of maximizing *EMV*.

A comment is in order on the probabilities of outcomes shown in the second column of the payoff table. Changing these probabilities may change the decision. But how good are these figures? Even though too little is known about the prospect to give exact values for the probability of finding oil or gas, the little information the investigator is able to give, like a general range of probabilities, may still help in deciding which course of action to take. This is illustrated by an example.

Example 8.3. Investment in drilling a well is considered. Four decision alternatives are available to the company:

1. Drill with 100 percent working interest.

2. Drill with 50 percent partner.

TABLE 8.2
Expected Monetary Value Computation For a Drilling Prospect

Possible Outcomes	Probability Outcome Will Occur	Drill with 50% WI		Farm Out with ORI		Penalty with Back In	
		Conditional NPV Profit ($)	Expected NPV Profit ($)	Conditional NPV Profit ($)	Expected NPV Profit ($)	Conditional NPV Profit ($)	Expected NPV Profit ($)
Dry hole	0.30	− 40,000	− 12,000	0	0	0	0
200 Mbbl	0.25	+ 50,000	+ 12,500	+ 12,000	+ 3,000	+ 12,000	+ 3,000
400 Mbbl	0.25	+300,000	+ 75,000	+ 60,000	+15,000	+145,000	+ 36,250
800 Mbbl	0.10	+700,000	+ 70,000	+120,000	+12,000	+400,000	+ 40,000
1,000 Mbbl	0.10	+800,000	+ 80,000	+130,000	+13,000	+500,000	+ 50,000
	1.00		EV = +225,500		EV = +43,000		EV = +129,250

3. Farm out, but back in for 50 percent working interest.

4. Do not drill.

What course of action should the company take considering the payoff table given below?

NPV Profit of Courses of Action ($1000)

Possible Outcomes	Drill with 100% *WI*	Drill with 50% Partner	Farm Out, Back In for 50% *WI*	Don't Drill
Dry hole (failure)	−500	−250	0	0
Producer (success)	3,000	1,500	1,500	0

Solution

Let P_s be the chance of success and NPV_s the net present value profit in case of success, and $(1 - P_s)$ be the chance of failure (a dry hole) with a net present value profit of NPV_f. The expected monetary value is

$$EMV = P_s NPV_s + (1 - P_s)NPV_f$$

$$= P_s(NPV_s - NPV_f) + NPV_f \qquad (8.1)$$

Thus EMV is a linear function of the probability of success, P_s. The corresponding forms of Eq. 8.1 for the four alternatives are:

Action 1   Drill with 100% *WI*: $EMV_1 = 3500P_s - 500$

Action 2   Drill with 50% Partner: $EMV_2 = 1750P_s - 250$

Action 3   Farm out, back in for 50% *WI*: $EMV_3 = 1500P_s$

Action 4   Do not Drill: $EMV_4 = 0$

These straight-line relationships are plotted in Fig. 8.3.

In this example if the investigator was not certain of the probability of finding oil reserves but felt that the chance was in the range of 0.6 to 0.7, the preferred choice would be Action 1 (i.e., drilling alone with 100 percent working interest). In fact, if the chance of success is 0.25 or more, Action 1 is advisable over Actions 2, 3, or 4. If the chance of finding oil reserves is less than 0.25, Action 3 (that is farm out and back in for 50 percent working interest) is advisable.

Graphs like Fig. 8.3 are very useful in exploration risk analysis. Even though the decision maker does not know the exact probability of finding oil or gas reserves, a firm and correct course of action can still be taken with a knowledge of the probable range of the probability of success.

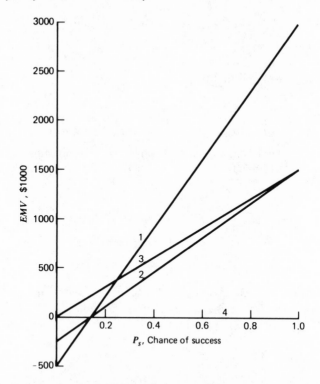

Figure 8.3   Relation between *EMV* and chance of success for the courses of action.

8.3   UTILITY THEORY

Faults of Expected Monetary Value (EMV)

Implicit in traditional analysis is the assumption that every dollar in the cash flow is equally important, except for a time-adjustment factor. This is clearly an oversimplification. Furthermore, it is not dollar consequences that are meaningful in establishing relative project worth, but rather it is the net utility of the project to the investor that is truly relevant.

Most formal analyses of business decisions involving uncertainty assume that every individual or every company has (or ought to have) the same attitude toward risk. The underlying assumption is that a decision maker will want to choose the course of action that has the highest expected value of profit (maximizing of *EMV*). The analyses usually assume that decision makers will want to "play the average" on all deals, regardless of the potential negative consequences that might result. But in fact, as every experienced executive knows, very few businessmen take this attitude toward risk when they make important decisions. The disadvantages of the *EMV* include:

1.   Failure to take into account the different financial circumstances of investors. Can they afford the loss (gambler's ruin)?

2. It is purely mechanical and fails to incorporate the judgment and experience of a successful decision maker.

3. It tends to place a more realistic value on alternative acts.

This is not to imply that *EMV* is not a valid decision parameter. It is completely valid if the decision maker is impartial to money. But very few (if any) decision makers are completely impartial to money. Most have specific attitudes and feelings about money that are caused by such factors as asset position, risk preferences, and goals. One choice left to the decision maker is to use *EMV* as an indicator of relative value and include consideration of his or her attitudes and feelings about the money involved in an informal, nonquantitative manner. A second choice is to incorporate his attitudes and feelings into a quantitative decision parameter having all the characteristics of the expected value concept and use the resulting numerical parameter as a basis for decision making under uncertainty. This second choice has led to the cardinal utility theory, also known as preference theory, first proposed by von Neumann and Morgenstern.

Utility Concepts

Utility theory states that each individual has a measurable preference when faced with choices among alternatives uncertainty, which is called his "utility" and is measured in terms of "units" (called utiles), the scale of which is set arbitrarily. The relationship between utiles and dollars is called an individual's utility function (curve). Utility theory also assumes that the utility function is predictive: an individual continues to apply the same utility function in future decision-making situations, given that conditions do not change radically. The function is strictly personal and subjective; it differs among individuals.

The recommended procedure for determining a person's utility function is to find a "point of indifference" when the decision maker has no preference among several alternatives and accepts them as possessing the same utility. This point is determined by questioning the decision maker about his preference when confronted with two monetary (or other) alternatives. The number of utiles fitting the "point of indifference" represents the person's utility function. Utility theory is a refinement of methods that assume the desirability of optimizing expected dollar income in risk decisions. It suggests that a man will (or perhaps, should) attempt to optimize expected utility rather than dollar gain—and that a function, relating utility to dollar gain, can be found for each rational individual.

Newendorp has enumerated the following important properties of the utility or preference curve (see Figs. 8.4 and 8.5):

1. The utility scale is dimensionless and reflects only a relative desirability of an amount of money. The magnitude of the scale is arbitrary until two points on the curve have been defined. The zero point on the vertical scale is generally interpreted as a point of indifference, or neutrality about money. Positive, upper values on the vertical axis denote increasing desirability and the negative portion of the scale denotes an increasing dislike for the corresponding amounts of money.

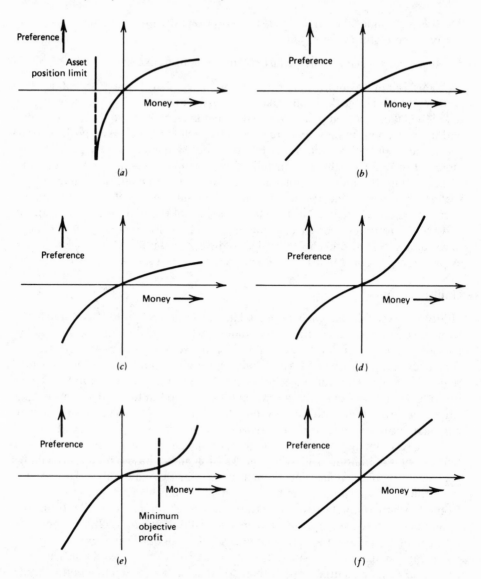

Figure 8.4 Shapes of various preference curves: suggest limited assets or capital constraint *(a)*, no limit in the range of money *(b)*, a conservative attitude *(c)*, a risk-seeking attitude *(d)*, an interest in profits only above a certain level *(e)*, and an interest that is totally impartial to money *(f)* (after Newendorp 1975).

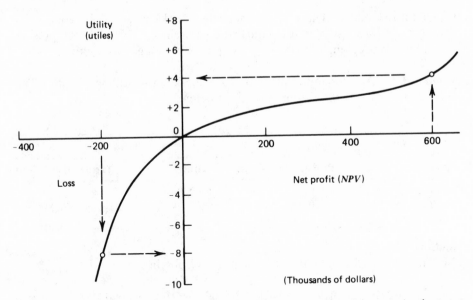

Figure 8.5 A typical utility curve for a company shows its risk preferences. A reluctance to participate in expensive ventures is indicated by the steep drop for losses greater than $100,000 (after J. M. Campbell).

2. The horizontal scale can be in any units of money, such as undiscounted monies, net present value monies discounted at a specified interest rate, incremental cash flow, and current asset position.

3. The curve is a monotonically increasing function.

4. The utility curve is based on an individual's preferences and does not imply a comparison among individuals. The curve merely describes a person's preferences and attitudes for money, and does not imply that he is wrong for having the attitude or that he should change his attitude about money.

5. The shape of the utility curve reflects the attitudes and preferences of the decision maker. If he was totally impartial to money (the assumption with *EMV*), his utility curve would be a straight line passing through the origin.

6. Utility theory has the property of expectation. We can compute expected utility value *(EUV)* for a decision alternative by multiplying the utility values and the probabilities of occurrence.

7. The *EUV* is the decision parameter used by the decision maker to accept or reject the alternative. The decision rule is to select the alternative that maximizes *EUV*.

Figure 8.4 illustrates the general shapes of utility curves. The shape in the first quadrant indicates the decision maker's attitudes toward risks. A utility curve that is concave downward in the first quadrant, as in Fig. 8.4*b*, portrays a conservative attitude for risk taking. This is sometimes called "risk-aversion." By contrast, a curve that bends sharply upward in the first quandrant as in Fig. 8.4*d* is representative of a "risk-seeking" attitude—the opposite of being conservative.

Example 8.4. A company is planning to embark on a drilling venture. The possible outcomes of the venture are given below.

Outcomes	Probability	Alternatives	
		Drill	Farm Out (1/8 *ORI*)
Dry hole	0.4	−$200,000	0
5 Bcf (Billion cubic feet)	0.6	+$600,000	+$50,000

Solution
Using the expected monetary value concept, we obtain

$$EMV \text{ of "Drill"} = (0.4)(-\$200,000) + (0.6)(+\$600,000)$$

$$= +\$280,000$$

$$EMV \text{ of "Farm out"} = (0.4)(0) + (0.6)(+\$50,000)$$

$$= +\$30,000$$

The *EMV* calculation shows that the alternative "drill" is the preferred choice.

The utility curve of this company is shown in Fig. 8.5. This curve indicates risk preferences. The curve drops rapidly for losses greater than $100,000, showing reluctance to participate in expensive ventures. Using the expected utility computation gives us

$$EUV \text{ of "Drill"} = (0.4)(-8) + (0.6)(+4)$$

$$= -0.80$$

$$EUV \text{ of "Farm out"} = (0.4)(0) + (0.6)(+0.6)$$

$$= +0.36$$

The utility computation shows that the preferred alternative is to "farm out."

The difference between the *EMV* and *EUV* decisions occurs because the potential loss of $200,000 overrides the potential gain of $600,000. Without the use of the utility theory, the company would have made the wrong decision!

Quantifying Utility

It is very difficult, if not impossible, to quantify utility functions to express the preference of a company for one type of project over another. Two methods that have been found useful in quantifying utility functions will be described.

Method I: Utilizing the Point of Indifference
A decision maker is faced with the choice of several alternatives. Let $U(x)$ be the utility of some choice x.

1. The decision maker is faced with the choice between $500,000 certain profit and 50–50 chance of $2,000,000. The decision maker chooses the $500,000 certain profit. The return on the second alternative is increased until the decision maker becomes indifferent at $4,000,000. Thus, to the decision maker the utility of $500,000 certain profit is equal to the utility of 50–50 chance of $4,000,000.

$$U(x_1) = U(x_2)$$

$$U(\$500,000) = 0.5U(\$4,000,000) + 0.5U(\$0)$$

The utility scale is dimensionless and reflects a relative desirability of an amount of money. Let $U(\$4,000,000) = 8$ utiles and $U(\$0) = 0$ utiles. Then,

$$U(\$500,000) = 0.5(8) + 0.5(0) = 4 \text{ utiles}$$

The three points $U(\$0)$, $U(\$500,000)$, and $U(\$4,000,000)$ can be plotted on the decision makers utility curve. To obtain a more complete curve, let us consider two more choices.

2. The decision maker has a small gas field that has a 50 percent chance of making $500,000 profit. The decision maker will not sell this property for $100,000; however, the point of indifference occurs at a selling price of $200,000. Then,

$$U(\$200,000) = 0.5U(\$500,000) + 0.5U(\$0)$$

or

$$U(\$200,000) = 0.5(4) + 0.5(0) = 2 \text{ utiles}$$

This point is plotted on the utility curve.

3. This example involves the purchase of a property. If the probability of $200,000 profit is 0.6 and the probability of $500,000 loss is 0.4, the decision maker will not be interested. The point of indifference has a 0.8 chance of a $200,000 profit and a 0.2 chance of a $500,000 loss. Then,

$$0.8U(\$200,000) + 0.2U(-\$500,000) = 0$$

or

$$U(-\$500,000) = -\frac{0.8}{0.2}U(\$200,000) = -8 \text{ utiles}$$

The complete utility curve for the decision maker is shown in Fig. 8.6.

Method II: Utilizing Maximum Loss
Another common and useful relationship for the utility function is

$$U(\text{profit}) = \frac{\text{profit}}{\text{profit} + \text{constant}} \qquad (8.2)$$

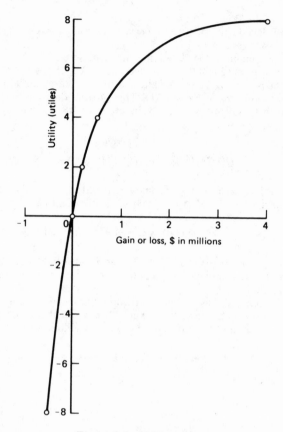

Figure 8.6 Utility curve.

It signifies that for every increase in profit, the increase in utility becomes less and less until at a very high value of profit the utility approaches the value 1. A zero profit gives a zero value of utility. For losses, utility becomes negative until at a loss equal to the constant, utility approaches minus infinity. This means that the constant in the equation is the maximum loss the investor can bear on his investment.

Consider two investors, A and B, using utility factors as functions of discounted profit as follows:

$$\text{Investor A: } U(\text{profit}) = \frac{\text{profit}}{\text{profit} + 2,000,000}$$

$$\text{Investor B: } U(\text{profit}) = \frac{\text{profit}}{\text{profit} + 8,000,000}$$

The utility curves are shown in Fig. 8.7.

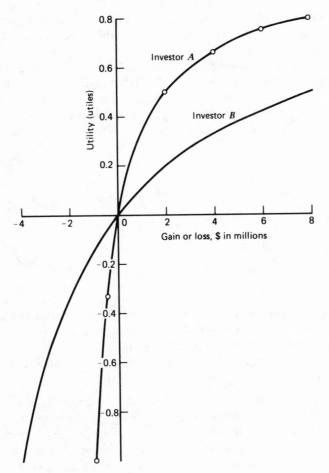

Figure 8.7 Utility curves of investors *A* and *B*.

8.4 DECISION TREES

Concepts

In a complex decision involving a long sequence of alternatives, a formal procedure for decision analysis is necessary to array the alternatives so that the economic ramifications of each are clearly delineated. This formal array also promotes effective internal communication. The decision tree analysis meets this criterion.

A decision tree is a pictorial representation of a sequence of events and possible outcomes. Since there is no scale to a decision tree, the lengths of the lines, or branches, have no significance. Also, since the angles between the branches have no meaning, the analyst need not worry about being precise in drawing the tree itself. The trees normally read from left to right and are drawn in the same order as the actual sequence in which the decision choices and chance events occur in the real world.

Some important definition on the decision tree is illustrated with a simple example. An investor is planning to buy a pizza restaurant near the university campus. If he buys the restaurant, there are three possible outcomes: a low demand, a medium demand, and a high demand for pizza. The probability of occurrence of each outcome and the discounted net present value profit are given below.

Possible Outcomes	Probability of Occurrence	NPV Profits
Low demand	0.6	−$ 25,000
Medium demand	0.3	+$ 50,000
High demand	0.1	+$150,000
	1.0	

1. The decision tree in Fig. 8.8 illustrates the decision alternatives "buy" and "don't buy," and the conditional outcomes of each decision. The decision maker dictates which branch is taken (buy or don't buy) but chance determines which of the three outcomes will occur if he elects to buy (low, medium, or high demand).

2. The point from which two or more branches emanate is called a node. A node surrounded by a square denotes a decision node, a point at which the decision maker dictates which branch is followed. An encircled node is called a chance node, a point where chance determines the outcome. Any number of decision alternatives or outcomes can emanate from a given decision or chance node. These nodes are indicated in Fig. 8.9.

3. For any set of outcomes resulting from chance, we associate probabilities of occurrence and monetary payoffs, as shown in Fig. 8.10. The decision tree drawn is interpreted as follows: there is a 60 percent chance of a low demand with a resulting loss of $25,000; a 30 percent chance of a medium demand and realizing a *NPV* profit of $50,000; and a 10 percent chance of a high demand and realizing a *NPV* profit of $150,000. The expected value of outcomes that might occur at the chance node is computed in the usual manner:

$$EMV = (0.6)(-\$25,000) + (0.3)(+\$50,000) + (0.1)(+\$150,000)$$
$$= +\$15,000$$

The sum of the probabilities of occurrence of outcomes emanating from a chance node must sum to one. The payoffs can be expressed in dollars, opportunity losses, present worth profits (or losses), or utility numbers (utiles).

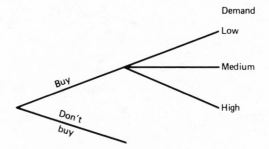

Figure 8.8 Partially completed decision tree.

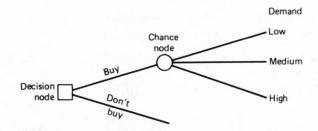

Figure 8.9 Decision tree with chance and decision nodes.

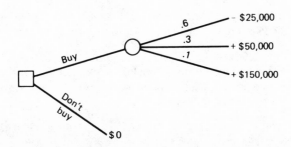

Figure 8.10 Example decision tree shows various options available, branching from either decision or chance nodes. Results of each possibility are given at the end of each branch.

4. The decision criterion at any decision node is to take the branch of alternative that maximizes expected value. For the simple illustration the alternatives are "buy" with *EMV* of $150,000 and "don't buy" with *EMV* of $0. The preferred alternative would be to buy the restaurant. Frequently the alternative not selected is crossed off on the decision tree as shown in Fig. 8.11.

5. The ends of a decision tree are called terminal points, indicating that there are no further contingencies, decisions, or chance events beyond that point. The terminal points of a decision tree are all said to be mutually exclusive;

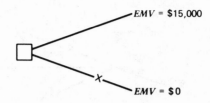

Figure 8.11   Decision maker's choices at decision node.

we will ultimately end up at one, and only one, terminal point on the tree. The uncertainty results because the decision maker cannot be sure at which of the terminal points he will end when he makes his initial decision.

Decision trees can be described according to their complexity. The following groups, except possibly for the last, differ more in degree than in kind:

a.   Probability trees, the ends of whose branches represent possible final outcomes of a venture, which will be determined by chance. These may be isolated or combined as parts of larger trees.

b.   A concise map that indicates the chronological order in which a series of actions will be performed. Major choices or decisions that must be made are indicated. The few branches presented are generally reduced to their expected values.

c.   A more detailed map whose branches indicate the range of possible outcomes and whose decision nodes are emphasized according to their future position values.

d.   An expanded tree with many branches for indicating various degrees of success. This becomes unwieldy unless programmed for a computer, as when we want to calculate expected values.

e.   A tree fabricated with selected probability distributions at the probability nodes and with decision rules built into the decision nodes. This becomes operable as a computer program whose output is a frequency distribution describing present-value variation. These, because of the high degree of interaction between successive outcomes and decisions, are referred to as stochastic decision trees.

The general procedure for constructing and using a decision tree is as follows:

1.   Identify the decision points and alternative actions available at each decision point.

2.   Identify the uncertainties (chance event points) and establish the type and range of alternative outcomes at each chance even point.

3.   Estimate the relevant quantitative information:

(a) Costs of the possible actions.

(b) Gains resulting from possible outcomes.

(c) Probabilities of chance events.

4. Define the criterion of desirability.

5. Evaluate the tree to obtain expectations and position values and to select the appropriate course of action.

Example 8.5.   The XYZ Enterprises has a nontransferable short-term option to drill on a certain plot of land. The option is the only business deal in which the firm is involved now or that it expects to consider between now and September 30, 1984, the time the drilling would be completed if the option is exercised. Two recent dry holes elsewhere have reduced XYZ's liquid assets to $130,000, and John Doe, president and principal stockholder, must decide whether XYZ should exercise its option or allow it to expire. It will expire in two weeks if drilling is not commenced by then. Doe has three possible choices:

1. Drill immediately.

2. Pay to have a seismic test run in the next few days and, then, depending on the result of the test, decide whether or not to drill.

3. Let the option expire.

XYZ can have the seismic test performed for a fee of $30,000, and the well can be drilled for $100,000. XYZ usually sells the rights of any oil discovered. A major oil company has promised to purchase all of the oil rights for $400,000.

The company's geologist has examined the available geological data and states that there is a .55 probability that if a well is sunk oil will be discovered. Data on the reliability of seismic tests indicate that if the test result is favorable, the probability of finding oil will increase to .85; but if the test result is unfavorable, it will fall to 0.10. The geologist has computed that there is a .60 probability that the result will be favorable if a test is made. This decision problem is structured in the form of a decision tree in Fig. 8.12.

Let's assume that Doe's decision rule is maximization of his expected asset position. At each terminal fork we proceed by computing the expected value of the firm's asset position. If we use the topmost terminal fork for illustration, the expected value is

$$(0.85)(+\$400,000) + (0.15)(\$0) = \$340,000$$

The expected value computations are presented in Fig. 8.13. The expected asset position of each event fork is shown enclosed in a box at the base of each event fork.

The two choices open to Doe in Fig. 8.14 are: "drill without seismic" for an expected net asset position of $250,000, or "drill with seismic" for an expected net asset position of $244,000. Doe should have his firm drill immediately without taking any seismic test.

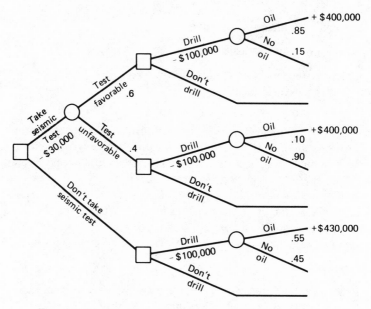

Figure 8.12 Doe's decision problem (XYZ enterprises).

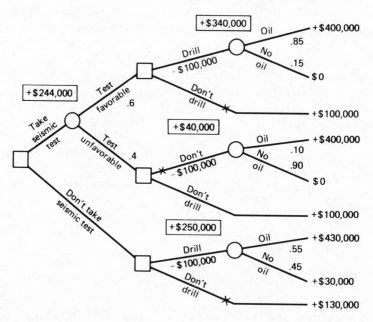

Figure 8.13 The decision tree for an example drilling venture shows options available and projected results. After computing *EMV*s at the three terminating chance nodes, obviously poor decisions are crossed off. Next, *EMV* at the first chance node is calculated using the remaining branches.

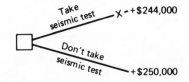

Figure 8.14 Doe's choices at decision node (XYZ Enterprises).

Comments

Consider John Doe's decision problem (Fig. 8.12). Instead of having a single *Oil* branch, we could split it into high, medium and low branches. Similarly, instead of splitting the *Take seismic* branch into only two branches, we might be more realistic by splitting it into a dozen branches.

Even though we cannot conveniently accommodate certain other realistic complications we have mentioned by putting more branches into an existing fork, it is possible that we can accommodate them by inserting additional forks in the body of the tree or by adding new branching points at the tips of the tree. Thus, for example, we could put in a chance fork that describes the economics of lifting the oil and another that describes the price of oil next year, the year after, and so on. Life goes on after the point shown on the tree, and it is somewhat arbitrary to cut off the analysis at any specific horizon date. But the farther we look ahead and the more refined our analysis becomes, the more complex the tree becomes and, if we carry matters to an extreme, the tree begins to resemble a gigantic bush. Remember the analysis requires us to assign probabilities to all chance branches and utilities to all consequences at the tips of the tree. What a brutal task!

In most realistic problems, in fact, one cannot possibly begin to chart out all the possible occurrences and choices far out into the foreseeable future. Compromises must be made; a touch of Art must be combined with Science. The following is the lament of an experienced decision analyst.

TREES—A Decision-Maker's Lament

I think that I shall never see
A decision complex as that tree—

A tree with roots in ancient days
(At least as old as Reverend Bayes);

A tree with trunk all gnarled and twisted
With axioms by Savage listed;

A tree with branches sprouting branches
And nodes declaring what the chance is;

A tree with flowers in its tresses
(Each flower made of blooming guesses);

A tree with utiles at its tips
(Values gleaned from puzzled lips);

A tree with stems so deeply nested
Intuition's completely bested;

A tree with branches in a tangle
Impenetrable from any angle;

A tree that tried to tell us "should"
Although its essence was but "would";

A tree that did decision hold back
'Til calculation had it rolled back.

Decisions are reached by fools like me,
But it took a consultant to make that tree.

> Michael H. Rothkopf
> *with apologies to Joyce Kilmer and to*
> *competent, conscientious decision*
> *analysts everywhere*

8.5   PROBABILITY, MONETARY, EMV, AND EUV MAPS

Geologists and exploration managers appreciate the usefulness of contour maps for the display of various types of information, including geological features such as structure and facies variation. Furthermore, contour maps can be used to represent statistical measures applied to geological data. Similarly, contour maps also can be effectively used to represent relationships that incorporate information from both the geological and business sides. For example, if *EMV* tables can be constructed for prospects at specific localities, it follows that "*EMV* surfaces" can be calculated and represented over an area by means of contours. Thus, an ultimate product of an integrated decision system could be an expression of investment opportunities on a regional basis by means of maps. At least four kinds of maps could be used. Probability maps could be used to express the likelihood of outcomes of specific acts over the area. In turn, a series of monetary maps could express the financial consequences of particular outcomes stemming from a specific act. Finally, the information represented by both the probability maps and the monetary maps could be incorporated in *EMV* maps, leading in turn to expected utility maps. The following example is taken from an excellent treatment of the subject by Harbaugh et al.

The concept of mapping a real surface by use of contour lines is readily extended to the mapping of imaginary surfaces. Since the probability of a particular event such as drilling a dry hole varies with respect to our interpretation of the geology, it is logical to express variations in probability with contour maps. Figure 8.15 illustrates a family of hypothetical probability surfaces. Assume that a wildcat well is to be located somewhere within the area represented by Figure 8.15. For this simple example, only four possible outcomes for the drilling of a well are considered: a dry hole, a discovery of 13,000 barrels of recoverable reserves, a 45,000 barrel discovery, and a 100,000 barrel discovery. In the real

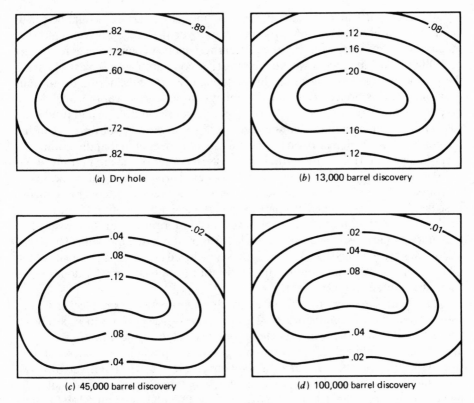

Figure 8.15 Series of four hypothetical probability maps that pertain to four mutually exclusive outcomes of wildcat well (after Harbaugh et al.)

world, of course, oil field magnitudes can be regarded as forming a continuum of possible sizes.

The sum of the values at a particular geographic location over the area must be 1.0, because there is absolute certainty of some outcome, and the outcomes have been defined as necessarily falling into one of four mutually exclusive classes. Thus, four probability surfaces are necessary to represent the variations over the area, and the surfaces form a complementary relationship so that their sum at any geographic location is 1.0.

The probability contours represented in Figure 8.15 must be regarded as probability estimates that are conditional on the present state of geologic knowledge. Assume that no wells have been drilled in the area. As soon as an exploration well has been drilled, new information becomes available, and the contours of the probabilities must be readjusted. If the well is dry, the probability of drilling another dry hole immediately adjacent to the existing dry hole is quite high (almost 1.0), and the probability estimates for the various magnitudes of success (*b, c,* and *d* of Fig. 8.15) necessarily has to be adjusted so that they are low in the immediate vicinity of the dry hole.

For each particular act and outcome, a "monetary surface" could be calculated and represented by contours of dollars over the area. For example, the outcome of a discovery would involve different monetary consequences for the act of drilling with 100 percent working interest, as compared with the act of having the prospect drilled as a farm out by someone else. By considering the various economic aspects, we can calculate the dollar consequences over the area. These may vary from place to place depending on differences in drilling and producing costs. For example, assume that the regional dip is toward the southwest, causing the potential producing horizons to become steadily deeper in that direction. The cost of a dry hole (Fig. 8.16) in the northeastern corner of the area is only about $25,000, but is about $135,000 in the southwestern corner. These differences will, of course, affect the dollar consequences in the event a discovery is made. Discoveries in the northeastern part of the area are more profitable (or involve smaller losses) than those in the southwest because of lower drilling costs. The monetary values also incorporate the effect of the discount rate. The large discoveries are assumed to take longer to be produced and, therefore, some of the oil to be produced in the comparatively distant future has a smaller dollar value when discounted to the present.

Monetary surfaces, as illustrated by the hypothetical examples of Figure 8.16, do not involve probabilities. Each monetary surface pertains to a specific act and a specific outcome. An expected monetary value surface, however, would involve the probabilistic representation of a series of possible outcomes resulting from a specific act. The value at each point on an *EMV* surface is derived from the probability estimates and the corresponding monetary consequences for all possible outcomes that stem from that particular act. Calculation of an *EMV* for each point thus involves multiplying a succession of probabilities by their monetary consequences and summing. Figure 8.17a provides an example of an *EMV* surface that represents drilling with a 100 percent working interest. It has been calculated by combining the information from Figs. 8.15 and 8.16 at a succession of points over the area. *EMV* surfaces for other possible acts could be computed in a similar manner.

The final step is to produce maps of expected utility for each particular act. Figure 8.17b illustrates an expected utility map for the act of drilling with 100 percent working interest for the hypothetical risk-averse utility function shown in Figure 8.18. The expected utility map is useful because it brings together virtually all relevant aspects of decision making. Geology is considered in the probabilities; exploration costs are considered in the expenses; gains from oil produced (if discovered) are included, and involve producing costs, royalties, taxes, a forecast of future oil prices, and a discount factor; finally, the operator's willingness to take risks versus his desire for gains are incorporated via his utility function.

Selection of the optimum investment in the hypothetical area of Figs. 8.15 to 8.17 is a matter of finding the largest expected utility value. This requires selection of a particular act to be taken at a particular location. If the desirable act at that location cannot be consummated (perhaps the land is already leased), the operator should then consider the location with the next highest expected utility, and so on. All acts considered should have positive utility.

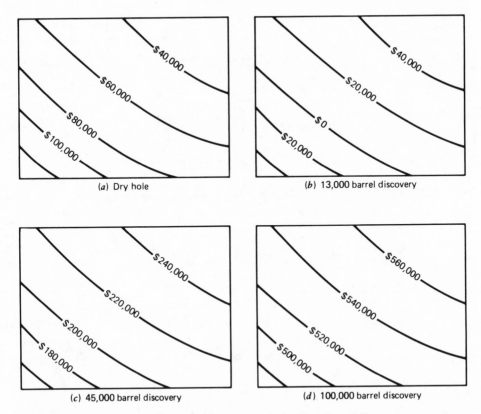

Figure 8.16 Series of maps of hypothetical area showing four different monetary outcomes for act of drilling single wildcat well with 100 percent working interest. Outcomes are mutually exclusive and represent all possible outcomes. Value of oil produced after deduction of royalty is $6.00 per barrel. Discount rate is ignored (after Harbaugh et al.)

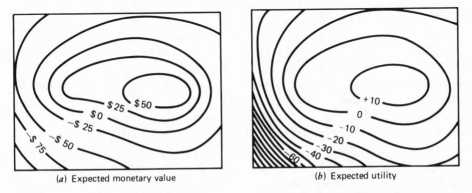

Figure 8.17 (a) *EMV* map in thousands of dollars for act of drilling with 100 percent working interest, (b) expected utility map of same area. Contours are in utiles. *EMVs* in map (a) have been transformed to utiles with utility function of fig. 8.18 (after Harbaugh et al.)

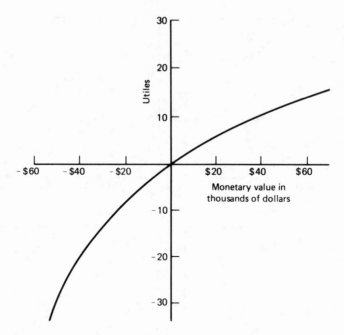

Figure 8.18 Utility function used in transforming *EMV* surface of fig. 8.17*a* to expected utility surface of fig. 8.17*b* (after Harbaugh et al.)

8.6   RISK AND UNCERTAINTY—MONTE CARLO SIMULATION

Perhaps no industry is more vitally concerned with risk and uncertainty than the oil and gas industry. Drilling decisions involving high investments are made very often. Even when an oil trap has been located, it is still difficult to tell the size of the field and whether it will pay off, before actual development drilling. Exploration for oil and gas therefore involves many uncertainties. Each elemental assumption in the exploratory search involves its own degree of uncertainty. Together these assumptions pyramid to a total uncertainty of critical proportions.

The old approach of using single-valued parameters while dealing with risk and uncertainty has undergone significant changes. The usefulness of stochastic processes is now being recognized. The probability that a variable has a certain value but may vary between certain limits often arises in oil and gas evaluations. Nature, using well the deck of cards in her possession, responds to the search for hydrocarbons and mineral deposits in general, by signals that are partially random. Thus, the determination of the size of an oil field lends itself easily to the Monte Carlo simulation technique.

The Monte Carlo technique stemmed from a concept proposed by von Neumann and Ulam in 1944. The von Neumann–Ulam concept suggested that many relationships arising in nonprobability contexts can be evaluated more easily by stochastic experiments than by standard analytical methods.

Probability Distributions

So far only discrete alternatives and their associated probabilities have been considered. Uncertainty in an estimate or forecast can also be represented by a probability distribution that expresses subjectively the relative chances that the variable estimated will turn out to have various different values. Some common shapes describing different types of uncertainty are shown in Fig. 8.19. Whereas for discrete alternatives the sum of the probabilities of all possible alternative must be 1, for continuous distributions the area under the distribution expresses the total probability of 1. The probability of the variable having a value in any particular range is the ratio of the area enclosed by that range to the total area under the distribution.

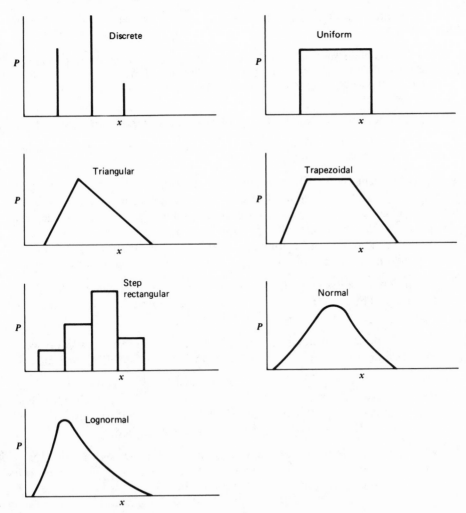

Figure 8.19 Some types of probability distributions *P* of a variable *x*. (courtesy of Institution of Chemical Engineers).

Some definitions are in order at this point.

Frequency is the number of occurrences, generally expressed as a fraction or percent of the total possible outcomes. It can be synonymous with probability.

Mean is the average of all possible outcomes.

Median is the middle-valued outcome.

Mode is the most likely or most probable outcome.

Variance is the mean square of the difference between the value of a variable and its mean or expected value.

Standard deviation is the square root of the variance.

For a symmetrical probability distribution, the mean, median, and mode are all the same.

> The Central Limit Theorem, *a basic law of statistics, can be stated: The distribution of the sums of random samples from any distribution with finite mean and variance approaches a standard normal (Gaussian) distribution as the individual sample size increases.*

The most important corollary of the Central Limit Theorem can be stated:

> *The distribution of the products of random samples approaches a lognormal distribution as the individual sample size increases.*

This theorem and its corrollary are important in the oil and gas industry in that porosity exhibits a standard normal distribution in a rock of consistent lithology. Permeability in a rock of consistent lithology and field reserve size in a single sedimentary basin exhibit lognormal distributions.

In the evaluation of the profitability of exploration and production ventures, most of the uncertainties are encountered in the following situations:

1. The number of discoveries (successes) resulting from drilling several exploratory wells. The chance of having x successes out of n trials is expressed by the binomial probability distribution.

2. The size (ultimate production) of the oil or gas field found by a discovery well. Field size distribution has been shown to approach a lognormal distribution.

3. The net present value profit per barrel (or *MCF*) of produced oil or gas. This distribution can be expressed by the uniform distribution if only the lower and upper limits of the *NPV* can be specified, or by the triangular distribution if the most likely value of the *NPV* can also be established.

1. ***The Binomial Distribution.*** If only two possible outcomes exist for a single event (such as success or failure), the probability of x successes in n trials is given by the binomial distribution function:

$$P_x = {_n}C_x\, p^x\, q^{n-x} \tag{8.3}$$

where

p = chance that a single event will be successful

q = chance that the event will fail ($p + q = 1$)

$_nC_x$ = the number of different ways x elements can be drawn from a total of n elements, disregarding the order in which they are obtained:

$$_nC_x = \frac{n!}{x!(n-x)!}$$

The mean of best estimate of the number of successes, $\mu = np$, and the standard deviation around μ is $\sigma_x = \sqrt{npq}$.

2. ***The Lognormal Distribution.*** The lognormal distribution is a continuous probability distribution that appears similar to a normal distribution except that it is skewed to one side (Fig. 8.20). Such a distribution means that not the property itself but the logarithm of the property is normally distributed. The cumulative distribution plotted on lognormal probability paper will result in a straight line if the distribution is lognormal. In nature a wide variety of properties are lognormally distributed, such as the family incomes in United States, the size of sand grains in a sediment, and the size of oil and gas fields resulting from past discoveries.

3. ***The Uniform and Triangular Distributions.*** The uniform and triangular distributions are believed to most reasonably approximate data encountered in oil and gas evaluations. These two distributions are commonly used in Monte Carlo simulation.

The *uniform distribution* confines the variable between an upper and a lower limit (Fig. 8.21). This distribution is used when no one range of values for a variable is more probable than any other, but when information or intuitive reasoning indicates that variables lie somewhere between the chosen limits. The

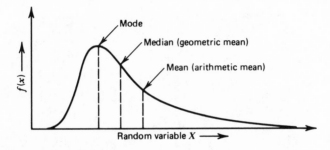

Figure 8.20   Lognormal distribution.

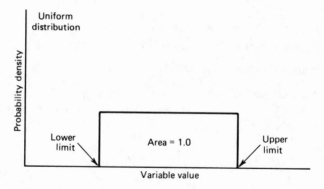

Figure 8.21 Uniform probability distribution confines the variable between an upper and lower limit when no range of values is more probable than another.

cumulative probability distribution is shown in Fig. 8.22. The mathematical expressions for the probability rules can be derived as follows.

Let

$$x_1 = \text{minimum value}$$

$$x_2 = \text{maximum value}$$

$$R_N = \text{uniformly distributed random number}$$

$$x = \text{required value of property}$$

Referring to Fig. 8.22, we get

$$x_1 \leqslant x \leqslant x_2$$

$$\text{Area} = 1 = y(x_2 - x_1)$$

$$f(x) = y = \frac{1}{x_2 - x_1}$$

Cumulative probability, $F = \int_{x_1}^{x} f(x)\, dx$

$$F = \int_{x_1}^{x} \frac{dx}{x_2 - x_1} = \frac{x - x_1}{x_2 - x_1}$$

whence

$$x = x_1 + F(x_2 - x_1)$$

Replace F by R_N, a uniformly distributed random number, where $0 \leqslant R_N \leqslant 1.0$.

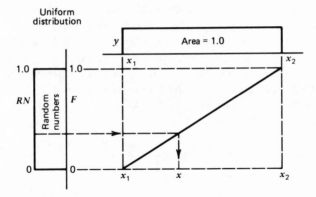

Figure 8.22 Uniform distribution showing variables used in mathematical model.

$$x = x_1 + R_N(x_2 - x_1)$$

The mean of the uniform distribution is $\mu = (x_1 + x_2)/2$ and the standard deviation is

$$\sigma = \sqrt{\frac{(x_1 - x_2)^2}{12}}$$

The *triangular distribution* is used for a variable when more data are available to indicate a central tendency of distribution (Fig. 8.23). This allows postulating a "most likely" value to the distribution, and upper and lower limits. In this case, as for the uniform distribution, the variable is not expected to assume a value less than the lower limit or greater than the upper limit. However, with improved quality of data it can be postulated that the variable tends to assume a value close to the most likely value, and that there is a decreasing probability for values away from the most likely value. The most likely value is not necessarily midway between the upper and lower limits.

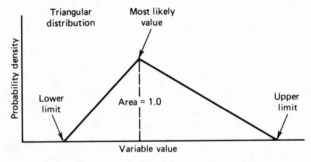

Figure 8.23 The triangular distribution is used when the variable is expected to assume a most likely value.

The total area under either of these probability distributions is equal to unity, since it is assumed that there is 100 percent probability that the variable lies somewhere under the curve. The probability is zero that the variable has any specific deterministic value. If two ordinates are drawn for any two values of the variable, the probability that the variable has a value lying between these ordinates is equal to the area under the curve lying between the ordinates. The cumulative probability distribution is shown in Fig. 8.24.

Referring to Fig. 8.24, we get

$$P_1 = \frac{x_2 - x_1}{x_3 - x_1}$$

$$P_2 = 1 - P_1 = 1 - \frac{x_2 - x_1}{x_3 - x_1} = \frac{x_3 - x_2}{x_3 - x_1}$$

where

$$x_1 = \text{minimum value}$$

$$x_2 = \text{most likely value}$$

$$x_3 = \text{maximum value}$$

Two cases will be considered:

Case 1: $x_1 \leqslant x \leqslant x_2$

$$\text{Area} = \frac{h}{2}(x_3 - x_1) = 1$$

$$h = \frac{2}{x_3 - x_1}$$

$$\frac{y}{x - x_1} = \frac{h}{x_2 - x_1}$$

$$y = h\left(\frac{x - x_1}{x_2 - x_1}\right) = \frac{2}{x_3 - x_1}\left(\frac{x - x_1}{x_2 - x_1}\right) = f(x)$$

Cumulative probability, $F = \int_{x_1}^{x} f(x)\, dx$.

$$F = \int_{x_1}^{x} \frac{2}{x_3 - x_1}\left(\frac{x - x_1}{x_2 - x_1}\right) dx = \frac{2}{(x_3 - x_1)(x_2 - x_1)}\left[\frac{x^2}{2} - xx_1\right]_{x_1}^{x}$$

$$= \frac{2}{(x_3 - x_1)(x_2 - x_1)}\left[\frac{x^2 - x_1^2}{2} - xx_1 + x_1^2\right] = \frac{(x - x_1)^2}{(x_2 - x_1)(x_3 - x_1)}$$

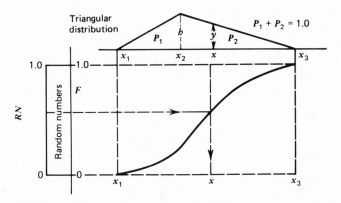

Figure 8.24 Triangular distribution showing variables used in mathematic model.

Replacing F with R_N, we get

$$R_N = \frac{(x - x_1)^2}{(x_2 - x_1)(x_3 - x_1)} \qquad (8.4)$$

and

$$x = x_1 + \sqrt{R_N(x_2 - x_1)(x_3 - x_1)} \qquad (8.5)$$

or

$$x = x_1 + (x_3 - x_1) \sqrt{R_N \frac{(x_2 - x_1)}{(x_3 - x_1)}} \qquad (8.6)$$

or

$$x = x_1 + (x_3 - x_1) \sqrt{R_N P_1} \qquad (8.7)$$

or

$$x = x_1 + (x_2 - x_1) \sqrt{\frac{R_N}{P_1}} \qquad (8.8)$$

Case 2: $x_2 \leqslant x \leqslant x_3$
By symmetry and replacing R_N by $(1 - R_N)$, we get

$$x = x_3 - \sqrt{(1 - R_N)(x_3 - x_2)(x_3 - x_1)} \qquad (8.9)$$

or

$$x = x_3 - (x_3 - x_1) \sqrt{(1 - R_N) \left(\frac{x_3 - x_2}{x_3 - x_1}\right)} \qquad (8.10)$$

or

$$x = x_3 - (x_3 - x_1) \sqrt{(1 - R_N)P_2} \qquad (8.11)$$

or

$$x = x_3 - (x_3 - x_2) \sqrt{\frac{(1 - R_N)}{P_2}} \qquad (8.12)$$

To select the appropriate case, use the following test criteria:

$$\text{If} \left(R_N - \frac{x_2 - x_1}{x_3 - x_1}\right) < 0, 0, > 0, \qquad \text{use Case 1, Case 1, Case 2}$$

or

$$\text{If} \ (R_N - P_1) < 0, 0, > 0, \qquad \text{use Case 1, Case 1, Case 2}$$

It can be proven that the best estimate of x is $\mu = (x_1 + x_2 + x_3)/3$. The standard deviation is given by

$$\sigma = \sqrt{\frac{(x_3 - x_1)(x_3^2 - x_1x_3 + x_1^2) - x_2x_3(x_3 - x_2) - x_1x_2(x_2 - x_1)}{18(x_3 - x_1)}}$$

A disadvantage of the triangular distribution is the sudden changes of the value of $f(x)$ at the points of minimum, most likely, and maximum values.

Simulation Technique

The basic idea of the technique is to carry out a large number of project evaluations with different input data selected from their specified distributions in random combinations. This is done in such a way that the frequency with which any value is selected corresponds to its probability in the distribution. After a large number of evaluations has been carried out, the result is a list of values of the profit indicator (e.g., *DCFROR*, and *NPV*). This list can be presented in the form of a frequency distribution of the profit indicator. The method depends on being able to select different values from a probability distribution with frequencies corresponding to their individual probabilities. This is done with the aid of random numbers.

Consider the discrete probability distribution in Fig. 8.25, in which the probabilities of *A*, *B*, and *C* are 0.3, 0.5, and 0.2, respectively. A properly balanced spinning disk is a simple random number generator. The circumference of the disk is divided into 10 equal parts (Fig. 8.26). The disk is spun beside a fixed marker and a random number is selected by noting the number nearest the marker when the disk comes to rest. To ensure that *A*, *B*, and *C* are selected with frequencies corresponding to their probabilities, the numbers on the disk are grouped to represent the probabilities of *A*, *B*, and *C* as in Fig. 8.27.

Monte Carlo simulation is usually programmed on a computer, since it involves a large number of trials. The numbers generated by a computer are not true random numbers but "pseudo" random numbers. To select values from a distribution, a computer program matches random numbers against the cumulative probability distribution (Fig. 8.28).

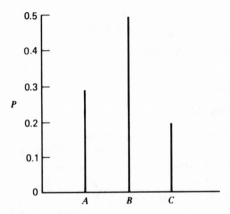

Figure 8.25 Discrete probabilities of *A*, *B* and *C*. (courtesy of Institution of Chemical Engineers).

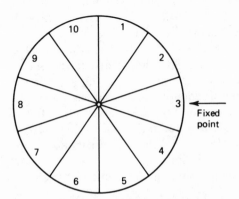

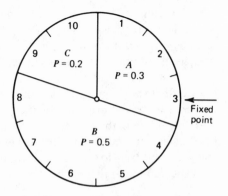

Figure 8.26 Spinning disc generator for 10 random numbers. (courtesy of Institution of Chemical Engineers).

Figure 8.27 Allocation of random numbers to select *A*, *B*, and *C*. (courtesy of Instittution of Chemical Engineers).

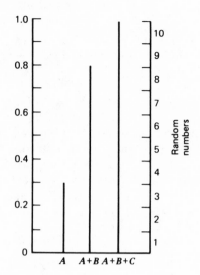

Figure 8.28 Matching random numbers against cumulative distribution. (courtesy of Institution of Chemical Engineers).

Economic Risk Analysis Model

In a classic paper in 1964, Hertz publicized the use of Monte Carlo simulation in investment analysis. Subjective probabilistic distributions are estimated for each of the input parameters. According to the probabilities they have of occurring in the future, point estimates are assigned to each variable. These selected values are then used in a valuation model to calculate the rate of return and/or present value of the investment. Figure 8.29 shows the procedure used by Hertz to implement the method.

The Monte Carlo simulation procedure applied to petroleum project evaluation is basically the same. The general logic of the method is summarized in Fig. 8.30. Distributions are defined for each random variable. For each pass through the evaluation corresponding to one potential project future, a random number is selected for each stochastic input and is matched with it to select a particular value from its distribution. These selected values along with other fixed data are used to calculate a value for the profit indicator, which is stored. This sampling scheme ensures that values selected over the series of repetitive computations are distributed exactly as the original specified distribution for the variable. All of this is normally done on the computer. When sufficient iterations (usually 100 to 1000) have been performed, the stored values for the resulting profit indicator are analyzed and transformed into a frequency distribution.

The economic risk analysis model should have the following features:

1. A main control program that specifies the number of Monte Carlo iterations and other controls and organizes the input data and the program as a whole.

2. A routine converting the input data to cumulative probability distributions.

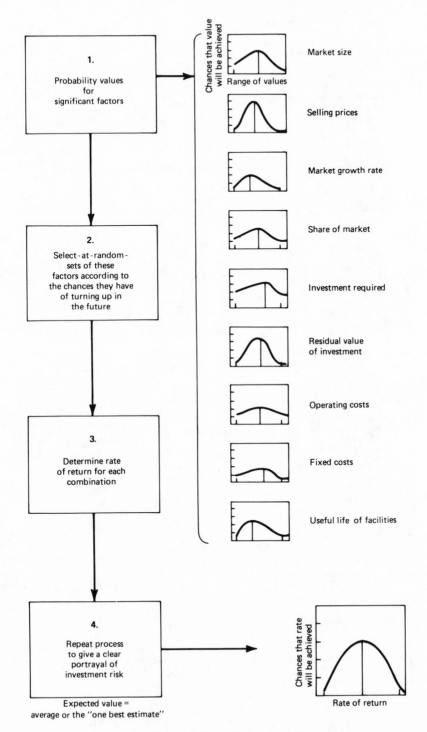

Figure 8.29 Hertz's Monte Carlo simulation model.

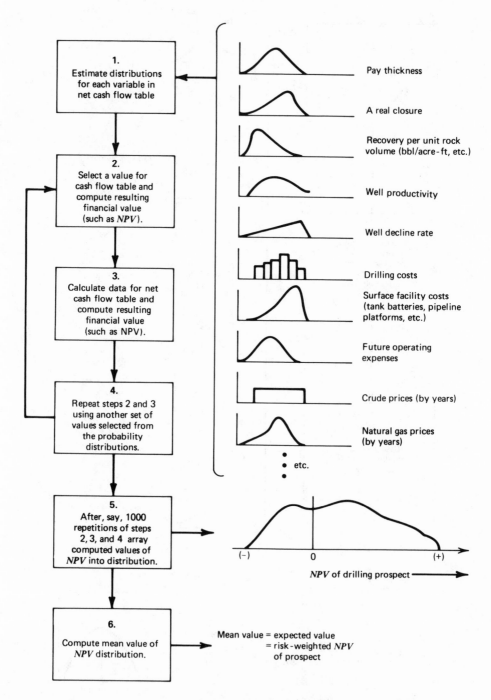

Figure 8.30   Exploration risk analysis model (after Newendorp 1975).

3. A routine that generates random numbers, matches them against the cumulative distributions, and selects values from the distribution for each iteration.

4. A routine that computes a set of annual cash flows for the project for each iteration.

5. Routines that compute profit indicators like *NPV* or *DCFROR* for each set of annual cash flows generated.

6. A routine analyzing the profit indicators (*NPV*s or *DCFROR*s) for all iterations and converts them into a frequency distribution as the final program output.

Field Size Distribution Using Monte Carlo Simulation

Generally, weighted averages are used to express reservoir properties. Deterministic solutions require controversial assumptions regarding the types of averages to be used in calculations. The Monte Carlo method circumvents the need for averaging, as it samples directly from distributions. In addition, the technique treats the solutions with continuing cognizance of the limiting ranges of accuracy and reliability.

The Monte Carlo simulation technique consists of mathematically simulating an experiment for a complicated expression involving one or more parameters, each of which has its associated uncertainty. The experiment uses a random sampling of the input parameter distributions involved in the expression being studied. Basically the method is one of numerical integration. The Monte Carlo method thus combines sampling theory and numerical analysis. A single run is synonymous with an experiment, and the output constitutes an observation. If a sufficiently large number of observations is averaged, the integrated outcome represents the solution that will be expected in the long run.

To determine oil in place, we consider porosity, area, formation thickness, water saturation, and formation volume factor as independent variables. A value of each independent random variable appearing in the expression is selected from its respective probability distribution. This set of values is then substituted into the expression,

$$N = \frac{7758 \; \phi \; Ah(1 - S_w)}{B_o}$$

where

$$7758 = \text{conversion factor}$$

$$\phi = \text{porosity, fraction}$$

$$A = \text{area, acres}$$

$$h = \text{thickness, feet}$$

S_w = water saturation, fraction

B_o = formation volume factor, vol/vol

N = STB of oil in place

A value of the dependent variable, N, is calculated. Subsequent values of oil in place are obtained by repeating the simulation process with additional sets of randomly sampled values of the five independent variables.

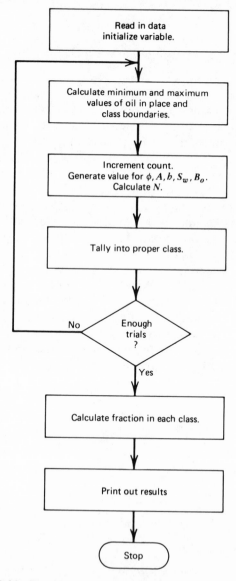

Figure 8.31 Flowchart for Monte Carlo simulation program.

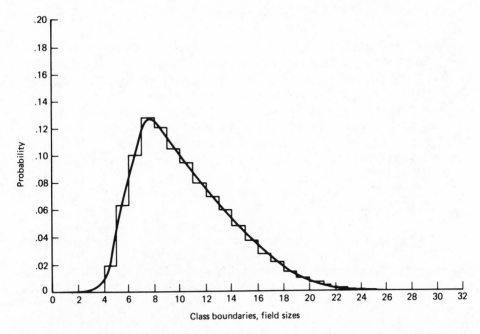

Figure 8.32   Results of a Monte Carlo simulation to determine field size distribution are skewed to the right indicating oil field sizes are lognormally distributed.

The flowchart for a typical Monte Carlo computer program used to determine the probability distribution of oil field size is presented in Fig. 8.31. The data were triangular distributed as follows:

	Minimum	Most Likely	Maximum
Porosity	0.12	0.20	0.25
Area	1500	2000	3000
Formation thickness	50	100	200
Water saturation	0.10	0.15	0.20
Formation volume factor	1.20	1.30	1.35

Five thousand (5000) Monte Carlo trials were made and the results were summarized in a probability distribution and plotted in Fig. 8.32. The distribution is skewed to the right and confirms that oil field sizes are lognormally distributed.

In a new exploration area it is important to be able to estimate the distribution of recoverable reserves per field in order to assess the feasibility of drilling. In this context we often speak of reserve distributions as being *lognormal*, and studies of these new areas frequently involve working with lognormal field size distributions.

REFERENCES

Allen, D. H.: *A Guide to the Economic Evaluation of Projects*, Institute of Chemical Engineers, Rugby, England, 1980.

Brown, R. V., Kahr, A. S., and Peterson, C.: *Decision Analysis for the Manager*, Holt, Rinehart and Winston, 1974.

Campbell, J. M.: "Decision Theory—Its Problems, Use, and Future," presented at Oklahoma City Section of SPE of AIME, December 19, 1967.

Campbell, J. M.: "Decision Trees—How to Formulate and Solve Sequential Decision Problems," presented at Oklahoma City Section of SPE of AIME, September 23, 1969.

Campbell, W. M.: "Risk Analysis: Over-all Chance of Success Related to Number of Ventures," *AIME*, 1961.

Cowan, J. V.: "Risk Analysis as Applied to Drilling and Developing an Exploration Prospect," paper SPE 2583, presented at 44th SPE Annual Fall Meeting, Denver, Colorado, September 28–October 1, 1969.

Davis, L. F.: "Economic Judgement and Planning in North American Petroleum Exploration," Journal of Petroleum Technology, May 1968, pp. 467–474.

Emery, D. R. and Tuggle, F. D.: "On the Evaluation of Decisions," *MSU Business Topics*, Spring 1976, p. 42.

Fleischer, G. A. and Cremer, R. H.: "On the Application of Cardinal Utility Theory to Engineering Economic Analysis," *The Engineering Economist*, Vol. 16, No. 2, January–February 1971, p. 117.

Grayson, C. J., Jr.: *Decision under Uncertainty—Drilling Decisions by Oil and Gas Operators*, Harvard University, 1960.

Gullatt, E. M.: "Decision Making—The State of the Art," presented at Oklahoma City Section of SPE and AIME, December 12, 1967.

Hammond, J. S.: "Better Decisions with Preference Theory," *Harvard Business Review*, November–December 1967, p. 123.

Harbaugh, J. W., Doveton, J. H., and Davis, J. C.: *Probability Methods in Oil Exploration*, John Wiley & Sons, New York, 1977.

Hax, A. C. and Wiig, K. M.: "The Use of Decision Analysis in Capital Investment Problems," *Sloan Management Review*, Winter 1976, p. 19.

Hertz, D. B.: "Risk Analysis in Capital Investment," *Harvard Business Review*, January–February 1964.

Ikoku, C. U.: "Decision Analysis: How to Make Risk Evaluations," Part I, *World Oil*, September 1980, p. 71.

Ikoku, C. U.: "Decision Analysis: How to Make Risk Evaluations," Part II, *World Oil*, October 1980, p. 157.

Kaufman, G. M.: "Statistical Analysis of the Size of the Oil and Gas Fields," SPE Paper No. 1096, presented at the 1965 Symposium on Petroleum Economics and Evaluation, Dallas, Texas, March 4–5, 1965.

Keeney, R. L.: "Examining Corporate Policy Using Multi-attribute Utility Analysis," *Sloan Management Review*, Fall 1975, p. 63.

Lahee, F. H.: "How Many Fields Really Pay Off?" *Oil and Gas Journal*, September 17, 1956, p. 369–371.

Magee, J. F.: "Decision Trees for Decision Making," *Harvard Business Review*, July–August, 1964.

Magee, J. F.: "How to Use Decision Trees in Capital Investment," *Harvard Business Review*, September–October 1964.

McCray, A. W.: *Petroleum Evaluation and Economic Decisions*, Prentice-Hall, 1975.

Newendorp, P. D.: *Decision Analysis for Petroleum Exploration*, Petroleum Publishing Company, 1975.

Newendorp, P. D.: "Mathematical Expectation—Expected Present Worth Profit," presented at Oklahoma City Section of SPE of AIME, September 16, 1969.

Raiffa, H.: *Decision Analysis, Introductory Lectures on Choices under Uncertainty*, Addison-Wesley, 1970.

Schlaifer, R.: *Analysis of Decisions under Uncertainty*, McGraw-Hill, 1969.

Smith, M. B.: "Estimate Reserves by Using Computer Simulations Method," *Oil and Gas Journal*, March 11, 1968, pp. 81–84.

Stoian, E.: "Fundamentals and Application of Monte Carlo Method," *Journal of Canadian Petroleum Technology*, July–September 1965, pp. 126–129.

Swalm, R. O.: "Utility Theory—Insights into Risk Taking," *Harvard Business Review*, November–December 1966.

von Neumann, J. and Morgenstern, O.: *Theory of Games and Economic Behavior*, Third Edition, Princeton, N.J., 1953.

Walstrom, J. E. et al.: "Evaluating Uncertainty in Engineering Calculations," *Journal of Petroleum Technology*, December 1967, pp. 1595–1603.

Wansbrough, R. S., Price, E. R., and Eppler, J. L.: "Evaluation of Dry Hole Probability Associated with Exploration Projects," paper SPE 6081, presented at SPE Annual Meeting, New Orleans, October 3–6, 1976.

Weinwurm, E. H.: "An Analysis of Applications of the Utility Concept," *The Engineering Economist*, Vol. 16, No. 2, January–February, 1971, p. 131.

Whipple, W., Jr.: "The Diutility of Uncertainty Losses," *The Engineering Economist*, Vol. 16, No. 2, October 1970, p. 43.

Wiest, J. D.: "Heuristic Programs for Decision Making," *Harvard Business Review*, September–October, 1966.

PROBLEMS

1. How many years are required for an investment to double in value with an interest rate of:
   a) 10% compounded annually?
   b) 10% compounded semi-annually?

2. For an annual interet rate of 10%:
   a) What is the present value of $4000 to be received 5 years hence, and $8000 to be received 10 years hence?
   b) How much will you pay me today for me to agree to pay back $1000 10 years from now?
   c) What investment in new equipment is justified if the new equipment will save me $1000 per year?
   d) How much is a $50,000 oil payment to be received 3½ years from now worth today?
   e) How many dollars to be received 6 years from now are equivalent to $15,000 to be received 12 years from now?

3. The net income for a project with a $45,000 investment is as follows:

Year	After-Tax Cash Flow
1	8,000
2	32,000
3	28,000
4	20,000
5	12,000

   Using mid-year convention,
   a) What is the present value of the project at a discount rate of 15%?
   b) What is the payout time?
   c) What is the profit per dollar invested?
   d) Calculate the discounted cash flow rate of return.

4. a) How much will $100 grow to if deposited at 4% for 85 years? What if the interest rate were 5% instead of 4%?
   b) One hundred dollars were deposited every year from 1890 to 1930. At 4%, what was this amount worth at the end of 1975?

c) How much can I pay now for a project will will yield $2000 each year at the end of the sixth, seventh, and eighth years, if money is worth 10% per year to me?

d) Take an initial investment of $300 that will have a yield of $50 per year for 4 years, plus an extra $200 at the end of the fourth year. What is the DCF rate of return?

5. You are an office manager with a monthly expense of $20,000 for your staff. You believe you can reduce this expense to $18,000 per month if you get a computer which will cost $100,000 to install, and will have a 5-year useful life. Is this a good reduction proposal? Assume no taxes for simplicity. Also, assume no salvage value.

6. Two alternative programs have been suggested to improve the returns of a program now in operation. You are asked to decide whether either of the alternatives or the present operation is the best investment choice.

Comparison of Investment Programs

Case	Inv. M$	Cash Flow Profit Each Yr, M$	NCR M$	P.O. Yrs.	DCF %
Base	2,000	1,000 for 8 yrs	6,000	2.0	62
Plan A	5,000	2,000 for 8 yrs	11,000	2.5	46
Plan B	11,000	3,000 for 8 yrs	13,000	3.7	26

Only one of these three programs can be chosen. Assume that risk is negligible, budget money is available, and your remaining inventory of budget opportunities averages 15%. For simplicity, assume no tax calculation.

7. The owner of a quarry signs a contract to sell her stone on the following basis: The purchaser is to remove the stone from certain portions of the pit according to a fixed schedule of volume, price, and time. The contract is to run 17 years, as follows: 7 years excavating a total of 20,000 yards per year at 10 cents per yard; the remaining 10 years excavating a total of 50,000 yards at 15 cents per yard. On the basis of equal year-end payments during each period by the purchaser, what is the present worth of the pit to the owner, if the interest rate is 5%.

8. Three systems are proposed for lowering the energy requirements of Jersey Hall. Available data indicate the following investment amounts and annual savings over the 30-year period selected for analysis.

Proposal	Initial Invesment	Energy Savings
R	$ 80,000	$12,000 per year first 10 years. $16,000 per year thereafter
S	$ 50,000	$ 8,000 per year
W	$100,000	$18,000 per year

No salvage value is assumed for any of the systems. The University of Tulsa requires a minimum rate of return of 15% before taxes on all cost savings investments.

a) What is the rate of return before taxes on each proposal?

b) Using an incremental rate of return approach which proposal should be adopted neglecting income taxes?

9. It is desired to determine whether to use no insulation or to use insulation 1 inch thick or 2 inches thick on a steam pipe in a liquefied natural gas complex. The heat loss from this pipe without insulation would cost $1.50 per year per foot of pipe. A 1-inch insulation will eliminate 89 per cent of the loss and will cost $0.40 per foot. A 2-inch insulation will eliminate 92 percent of the loss and will cost $0.85 per foot. Using an interest rate of 10 percent, compare the annual costs per 1000 feet of steam pipe with no insulation and with the two thicknesses of insulation, using a life of 10 years for the insulation with no salvage value.

10. A new gathering line must be constructed from an existing pumping station to an oil field 2000 ft distant. An estimate of costs for three pipe sizes has been made as follows:

Pipe Size inches	Cost per Hour for Pumping	Estimated Construction Cost
8	$2.00	$19,000
10	1.50	37,000
12	0.75	62,000

The pipe has a life of 16 years at the end of which there will be no salvage value. Interest is at 15% per annum.

a) How many hours of pumping per year would be required to make the 8-inch and 10-inch pipes equally economical?

b) How many hours of pumping per year would be required to make the 10-inch and 12-inch pipes equally economical?

c) For what ranges of required pumping hours per year is each size of pipe most economical?

11. Given an initial investment of $650,000 in a drilling fund and income of $800,000 obtained as shown below:

Year (end)	Annual Income (in $1000)
1	300
2	200
3	150
4	100
5	50
	800

Calculate:

    a) The pay-out time

    b) Profit-to-investment ratio

    c) Net present value profit at 15%

    d) Discounted profit-to-investment ratio at 15%.

Another investment proposal requires an initial investment of $1,000,000 and will provide an income of $2,240,000 in 20 years at a constant rate. Calcualte the same properties as above for this proposal. Compare the profitability of this investment with the drilling fund investment.

12. You own an oil property for which you paid $450,000 in mineral rights acquisition costs last year. Recoverable oil reserves are estimated at 1,000,000 barrels. 50,000 barrels of oil are produced this year and are sold for $30.00 per barrel. Your operating and overhead expenses are $240,000 this year and allowable depreciation is $60,000. You also expect the same production rate, operating costs, depreciation, and selling price in future years. Assuming that you are a small producer eligible for either cost or percentage depletion and that the selling price is after royalties, calculate the cash flow for three years. Income tax rate is 48%. Percentage depletion rate is 18%, 16%, and 15% for 1982, 1983, and 1984, respectively, and will remain at 1984 level thereafter.

13. Under existing conditions, you are going to receive a revenue of $10,000 in the tenth year from now (mid-year). Would you spend $3,000 today in order to accelerate this revenue to the fourth year from now (mid-year)?
    a) Set up the cash flow pattern for Case A, existing conditions.
    b) Set up the cash flow pattern for Case B, accelerated condition.
    c) Take the incremental cash flow, and discount to Present value over the range of 5% to 40%.
    d) Plot the Present Value Profile.
    e) Comment on the profitability of the acceleration project.

14. A well will produce a net income of $15,000 per year for 5 years and then go on exponential decline. The decline period will be 20 years and the ratio of the net income at abandonment to that initially is 0.01. For an annual interest rate of 15%, calculate (using mid-year convention)
    a) The present value of the allowable production.
    b) The value at the start of decline of the production during decline.
    c) The present value of the production during decline.
    d) The present value of the composite production.

15. An oil property with estimated reserves of 1,000,000 barrels of oil is leased for $500,000. 200,000 barrels of oil are pumped this year with annual operating costs of $60,000. Depreciation and amortization deductions total $150,000 this year. The oil is sold for $25.00 per barrel after royalties. Should cost depletion or percentage depletion be used for this year's tax deduction? What would be the 2nd year cost depletion?

16. Some mineral processing equipment has an initial value of $2,500 and a salvage value of $400 after a useful life of 6 years. The equipment is used for the following number of hours: first year, 2,000; second year, 1,800; third year, 1,000; fourth year, 600; fifth year, 400; sixth year, 200. Calculate the annual depreciation schedule using:
    a) Straight line.
    b) Double-Declining Balance (DDB).

c) Sum-of-Years Digits (SYD).

d) Units of Production.

17. a) A machine is purchased on January 1, 1985, for $4,200. Delivery charges are $200 and installation costs $600. Expected useful life of the equipment is 6 years, at which time it is anticipated that it can be dismantled at a cost of $200 and sold for $1,000. Compute the depreciation charge for the 4th year by each of the following depreciation methods:

   i) straight-line

   ii) sum-of-years digits

   b) What is the book value of the machine at the end of the 4th year using each of the depreciation methods?

   c) It is estimated that the machine of (a) has a capacity to produce 12,000 hours of production. Compute the depreciation charge for the 4th year, in which it produced 3,000 hours of work, by the unit of production method of depreciation.

   d) Stipulate a type of machine and situation in which it would be desirable to use each of the methods of (a) and (c).

   e) Why do companies generally prefer to depreciate their equipment more rapidly than permitted by income tax laws even though it results in the appearance of lower profits on their profit and loss statements in the early years?

18. An existing oil field will yield a total operating cash income of $5 million in 25 years at a constant rate. By infill drilling and increasing pump sizes the same income can be obtained in 10 years, also at a constant rate. The cost of this acceleration program is $1.25 million.

   For what company discount rates is this an attractive project? Use end of year discounting.

19. The federal income tax laws permit switching from declining balance or sum-of-the-years-digits method of depreciation to a straight-line rate after n years, where n will be governed by existing regulations. Under this procedure the undepreciated balance is subject to a straight-line rate over the remaining years of the asset's life. Consider investment in an oil production facility. $P = \$8000, L = \$2000, N = 5$ years. We will be allowed to use only the basic depreciation methods, without reducing the salvage value or useful life.

   Consider the possibility of switching from the sum-of-the-years-digits method to straight-line depreciation after either 1, 2, or 3 years assuming that the regulations will let us switch at any year that is to our benefit). From the point of view of minimizing the present worth of future income taxes, after which year should the switch be made if a corporation has a before-tax return of $10,000 each year? Use an interest rate of 12% in present worth determination and an income tax rate of 48 percent. Use end of year discounting.

20. Barbara is the sole owner of Y Corporation, in which she has personally invested $400,000. She has net taxable personal income of $500,000 per year from sources other than Y Corporation. Barbara has been offered $5,000,000 for the business, which she considers a fair price, but she decides not to sell. Y

Corporation shows an annual net profit of $2,000,000 before income taxes. There are no state or local income taxes.

a) What is the net profit after income taxes of Y Corporation?

b) If Y Corporation distributes these profits as dividends, how much of the $2,000,000 will Barbara retain after taxes? What has been the effective tax rate on the $2,000,000 profit? Use current individual income tax rates.

21. An oil producing project involves production of a 3 million barrel reserve. Acquisition and exploration costs of $2,000,000 have been capitalized into the cost depletion basis. Development expenses of $1,500,000 will be recovered ratable (unit of production depreciation) over the project life. Producing equipment costing $3,000,000 will be depreciated by DDB depreciation using a 10-yr life. Production is estimated to be 300,000 barrels per year at an average selling price of $30 per barrel after royalties, with operating costs of $3,000,000 per year. The effective income tax rate is 50%. Determine the net profit and cash flow each year for the first two years of project life, if this is the only oil or gas producing property owned by this company.

22. A lease containing four gas wells can produce at a maximum flow rate of 50 MMscfd because of limitation in the capacity of processing equipment. The wells will flow at this rate for about 8 years and then decline to an abandonment rate of 1 MMscfd after 17 years of declining production. Assume exponential decline. The net value of the gas is 10¢ per Mscf. Calculate the net present values of production at discount rates of 10% and 20%. If the total investment in the lease is $10,000, estimate the rate of return of the project.

23. Calculate the present value, before income taxes, of the following prospect. Use 15% discount factors; assume all receipts and expenditures over the year to occur at the midpoint of the year. Assume that exponential decline will follow after the period of restricted production. The basic date of evaluation is the beginning of the allowable production period, which is the date of well completion.

The data are as follows:

Well spacing:	40 acres
Net pay thickness:	26.5 ft
Recovery:	150 bbl/acre-ft
Initial flow capacity:	110 bbl/day
Maximum allowable production rate:	50 bbl/day
Working interest:	7/8 working interest
Cost of well:	$1,000,000
Operating cost, flowing (2 years):	$1,000 per month
Operating cost, pumping	$2,800 per month
Pumping equipment and installation:	$60,000

Assume 30 per barrel, less 5% state (severance) tax. From the working interest, allow 3% for ad valorem taxes and overhead.

24. Assume drilling costs are *uniformly distributed* and range between $40 and $50 per foot drilled.
    a) What is the most likely drilling cost?
    b) Forty percent of the drilling costs will be less than what value?
    c) Are you 80% sure the cost will fall between what two values?

25. Assume pipeline costs are *triangularly distributed* and range between $200,000 and $300,000 per mile with a most likely value of $240,000.
    a) Fifty percent of the pipeline costs will be less than what value?
    b) Are you 80% sure the cost will fall between what two values?
    c) Forty percent of pipeline cost values will be greater than what cost?

26. After doing *Monte Carlo ROR* analysis assume your answers are *normally distributed* with a mean of 40% and a standard deviation of 5%. What is the probability your actual ROR will be greater than 30%?

27. A new play in the Anadarko Basin will cost $1,000,000 per exploration well with a success rate of one producer per nine wells drilled. How many exploration wells must we invest in to be 85% sure of at least one successful well?

28. An oil lease of 160 acres has four wells already drilled on 40-acre spacing. The net pay is 25 ft. Assuming that ultimate recovery is discretely distributed with a low of 100 (prob = .2), most likely 200 (prob = .5) and high of 350 bbl/acre-ft. of pay. The four wells combined have already produced a total of 120,000 STB.
    Your company is considering buying this lease. If operating profit is considered a constant $25/STB, remaining reserves are produced at a constant rate over 10 years in all cases and one workover per well ($60,000 each) must be performed in year 4 (end of), what is the top figure you should pay for this oil production?

29. The figure on page 246 shows a wildcatter's decision tree. She, of course, would like to maximize the expected payoff. Recommend the course of action.

30. Suppose your true utility function for monetary gains or losses is given by

$$U(X) = 0.95X - 0.05X^2$$

where X is the gain or loss (in the range $-\$10$ to $+\$10$).
    a) Find your expected utility in a situation where you have a 0.6 probability of winning $8 and a 0.4 probability of losing $5.
    b) Show that you will always prefer not gambling to any gambling situation where you have a 50% chance of either winning or losing $10.
    c) This function has $U(-1) = -1$ and $U(0) = 0$. If you were faced with a gambling situation where you said you were indifferent between winning $5 and losing $1 at an indifference probability value of 0.30 (that probability of winning $5 = 0.30 and probability of losing $1 = 0.70) what is the difference between your true utility function value at $5 and the value based on your response?
    d) How would you qualitatively classify an individual that has the above utility function?

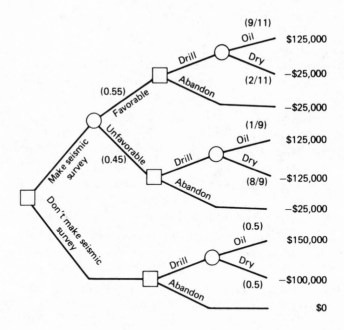

31. A drilling venture in an exploratory area has two possible outcomes: success S, representing a discounted profit of $5,000,000 and failure F, representing a loss of $100,000. Based on the best information presently available, the chance of success is $P(S) = 0.30$ and the chance of failure $P(F) = 0.70$.

a) If it is decided to drill, calculate the expected (discounted) profit based on this information. Any comments?

To investigate if we may improve the expected value calculated above by having some seismic work carried out, suppose this seismic survey may lead to two conclusions: either the results are favorable for success of drilling (g) or unfavorable (b). With this added information we like to know:

b) What is the chance of success $P(S/g)$, if the results of the survey are favorable (g)?

c) What is the chance of failure $P(F/g)$, if the results of the survey are favorable (g)?

d) What is the chance of success $P(S/b)$, if the results of the survey are unfavorable (b)?

e) What is the chance of failure $P(F/b)$, if the results of the survey are unfavorable (b)?

f) Compare the expected monetary value with seismic with the expected monetary value without seismic. Should seismic be carried out? If your answer is yes, how much money would you be willing to spend on the seimic. Why?

g) Sketch a decision tree for this problem.

Given: $p(g/S) = 0.8$; $p(b/S) = 0.2$
           $p(g/F) = 0.35$; $p(b/F) = 0.65$

h) Consider an investor with a utility function as follows:

$$u = \frac{\text{profit}}{\text{profit} + 1,000,000}$$

which means that the investor can bear a maximum loss of $1,000,000. Using expected utility value analysis, should Seismic be carried out or not?

32. An oil drilling company is considering the extent of drilling it should conduct on land where a large petroleum refiner has given it an exclusive drilling contract for a period of 2 years. The company will receive a payment from the refiner depending on the potential number of producing wells it discovers. It will select one of the following alternatives labelled A, B, C.
(A) engage in extensive drilling for both years,
(B) engage in light drilling the first year and make a decision after the first year on what to do the second year,
(C) engage in light drilling both years.
The investment costs (in millions of dollars) incurred at the beginning of each year for each alternative are:

Alternative	Year 1	Year 2
A	0.7	0.8
B	0.4	1.3 (if extensive drilling in year 2)
B	0.4	0.5 (if light drilling in year 2)
C	0.3	0.3

Success (S) or failure (F) in each year is determined by whether a minimum number of potential producing wells will be discovered. The company estimates that the chance of S or F in each year will primarily depend on the oil reserves on the land and not on the level of effort expended. However, the number of producers and hence the payment from the refiner will depend on both the effort and whether the reserves are there or not. It is also true that failure (or success) the first year does not guarantee failure (or success) the second year since there is also a chance that most of the drilling took place in the "wrong" places. However, any actual outcome in the first year would appear to enhance the chance of that outcome in the second year. Estimates for the chances of various combinations of outcomes are:

$S_1 S_2 = 0.45$; $S_1 F_2 = 0.20$; $F_1 S_2 = 0.10$; $F_1 F_2 = 0.25$.
(S_i is success in year i, F_i is failure in year i
$i = 1, 2$)

Payment to the driller for each year will be as shown below:

Extensive drilling and S	$1.5 million
Extensive drilling and F	$0.5 million

Light drilling and S	$0.7 million
Light drilling and F	$0.2 million

a) Draw a decision tree of the driller's problem.
b) What should be done to maximize his expected profit? Neglect time value of money.

Appendix A

DISCOUNT FACTORS FOR ANNUAL COMPOUNDING AND YEAR-END CASH FLOWS

				Annual Interest Rates			
Year	2%	3%	4%	5%	6%	7%	8%
1	0.98039	0.97087	0.96154	0.95238	0.94340	0.93458	0.92593
2	0.96117	0.94260	0.92456	0.90703	0.89000	0.87344	0.85734
3	0.94232	0.91514	0.88900	0.86384	0.83962	0.81630	0.79383
4	0.92385	0.88849	0.85480	0.82270	0.79209	0.76290	0.73503
5	0.90573	0.86261	0.82193	0.78353	0.74726	0.71299	0.68058
6	0.88797	0.83748	0.79031	0.74622	0.70496	0.66634	0.63017
7	0.87056	0.81309	0.75992	0.71068	0.66506	0.62275	0.58349
8	0.85349	0.78941	0.73069	0.67684	0.62741	0.58201	0.54027
9	0.83676	0.76642	0.70259	0.64461	0.59190	0.54393	0.50025
10	0.82035	0.74409	0.67556	0.61391	0.55839	0.50835	0.46319
11	0.80426	0.72242	0.64958	0.58468	0.52679	0.47509	0.42888
12	0.78849	0.70138	0.62460	0.55684	0.49697	0.44401	0.39711
13	0.77303	0.68095	0.60057	0.53032	0.46884	0.41496	0.36770
14	0.75788	0.66112	0.57748	0.50507	0.44230	0.38782	0.34046
15	0.74301	0.64186	0.55526	0.48102	0.41727	0.36245	0.31524
16	0.72845	0.62317	0.53391	0.45811	0.39365	0.33873	0.29189
17	0.71416	0.60502	0.51337	0.43630	0.37136	0.31657	0.27027
18	0.70016	0.58739	0.49363	0.41552	0.35034	0.29586	0.25025
19	0.68643	0.57029	0.47464	0.39573	0.33051	0.27651	0.23171
20	0.67297	0.55368	0.45639	0.37689	0.31180	0.25842	0.21455
21	0.65978	0.53755	0.43883	0.35894	0.29416	0.24151	0.19866
22	0.64684	0.52189	0.42196	0.34185	0.27751	0.22571	0.18394
23	0.63416	0.50669	0.40573	0.32557	0.26180	0.21095	0.17032
24	0.62172	0.49193	0.39012	0.31007	0.24698	0.19715	0.15770
25	0.60953	0.47761	0.37512	0.29530	0.23300	0.18425	0.14602
26	0.59758	0.46369	0.36069	0.28124	0.21981	0.17220	0.13520
27	0.58586	0.45019	0.34682	0.26785	0.20737	0.16093	0.12519
28	0.57437	0.43708	0.33348	0.25509	0.19563	0.15040	0.11591
29	0.56311	0.42435	0.32065	0.24295	0.18456	0.14056	0.10733
30	0.55207	0.41199	0.30832	0.23138	0.17411	0.13137	0.09938
35	0.50003	0.35538	0.25342	0.18129	0.13011	0.09366	0.06763
40	0.45289	0.30656	0.20829	0.14205	0.09722	0.06678	0.04603
45	0.41020	0.26444	0.17120	0.11130	0.07265	0.04761	0.03133
50	0.37153	0.22811	0.14071	0.08720	0.05429	0.03395	0.02132
60	0.30478	0.16973	0.09506	0.05354	0.03031	0.01726	0.00988
70	0.25003	0.12630	0.06422	0.03287	0.01693	0.00877	0.00457
80	0.20511	0.09398	0.04338	0.02018	0.00945	0.00446	0.00212
90	0.16826	0.06993	0.02931	0.01239	0.00528	0.00227	0.00098
100	0.13803	0.05203	0.01980	0.00760	0.00295	0.00115	0.00045

			Annual Interest Rates				
Year	9%	10%	11%	12%	13%	14%	15%
1	0.91743	0.90909	0.90090	0.89286	0.88496	0.87719	0.86957
2	0.84168	0.82645	0.81162	0.79719	0.78315	0.76947	0.75614
3	0.77218	0.75131	0.73119	0.71178	0.69305	0.67497	0.65752
4	0.70843	0.68301	0.65873	0.63552	0.61332	0.59208	0.57175
5	0.64993	0.62092	0.59345	0.56743	0.54276	0.51937	0.49718
6	0.59627	0.56447	0.53464	0.50663	0.48032	0.45559	0.43233
7	0.54703	0.51316	0.48166	0.45235	0.42506	0.39964	0.37594
8	0.50187	0.46651	0.43393	0.40388	0.37616	0.35056	0.32690
9	0.46043	0.42410	0.39092	0.36061	0.33288	0.30751	0.28426
10	0.42241	0.38554	0.35218	0.32197	0.29459	0.26974	0.24718
11	0.38753	0.35049	0.31728	0.28748	0.26070	0.23662	0.21494
12	0.35553	0.31863	0.28584	0.25668	0.23071	0.20756	0.18691
13	0.32618	0.28966	0.25751	0.22917	0.20416	0.18207	0.16253
14	0.29925	0.26333	0.23199	0.20462	0.18068	0.15971	0.14133
15	0.27454	0.23939	0.20900	0.18270	0.15989	0.14010	0.12289
16	0.25187	0.21763	0.18829	0.16312	0.14150	0.12289	0.10686
17	0.23107	0.19784	0.16963	0.14564	0.12522	0.10780	0.09293
18	0.21199	0.17986	0.15282	0.13004	0.11081	0.09456	0.08081
19	0.19449	0.16351	0.13768	0.11611	0.09806	0.08295	0.07027
20	0.17843	0.14864	0.12403	0.10367	0.08678	0.07276	0.06110
21	0.16370	0.13513	0.11174	0.09256	0.07680	0.06383	0.05313
22	0.15018	0.12285	0.10067	0.08264	0.06796	0.05599	0.04620
23	0.13778	0.11168	0.09069	0.07379	0.06014	0.04911	0.04017
24	0.12640	0.10153	0.08170	0.06588	0.05323	0.04308	0.03493
25	0.11597	0.09230	0.07361	0.05882	0.04710	0.03779	0.03038
26	0.10639	0.08391	0.06631	0.05252	0.04168	0.03315	0.02642
27	0.09761	0.07628	0.05974	0.04689	0.03689	0.02908	0.02297
28	0.08955	0.06934	0.05382	0.04187	0.03264	0.02551	0.01997
29	0.08215	0.06304	0.04849	0.03738	0.02889	0.02237	0.01737
30	0.07537	0.05731	0.04368	0.03338	0.02557	0.01963	0.01510
35	0.04899	0.03558	0.02592	0.01894	0.01388	0.01019	0.00751
40	0.03184	0.02209	0.01538	0.01075	0.00753	0.00529	0.00373
45	0.02069	0.01372	0.00913	0.00610	0.00409	0.00275	0.00186
50	0.01345	0.00852	0.00542	0.00346	0.00222	0.00143	0.00092
60	0.00568	0.00328	0.00191	0.00111	0.00065	0.00039	0.00023
70	0.00240	0.00127	0.00067	0.00036	0.00019	0.00010	0.00006
80	0.00101	0.00049	0.00024	0.00012	0.00006	0.00003	0.00001
90	0.00043	0.00019	0.00008	0.00004	0.00002	0.00001	0.00000
100	0.00018	0.00007	0.00003	0.00001	0.00000	0.00000	0.00000

Year	\ Annual Interest Rates						
	16%	17%	18%	19%	20%	21%	22%
1	0.86207	0.85470	0.84746	0.84034	0.83333	0.82645	0.81967
2	0.74316	0.73051	0.71818	0.70616	0.69444	0.68301	0.67186
3	0.64066	0.62437	0.60863	0.59342	0.57870	0.56447	0.55071
4	0.55229	0.53365	0.51579	0.49867	0.48225	0.46651	0.45140
5	0.47611	0.45611	0.43711	0.41905	0.40188	0.38554	0.37000
6	0.41044	0.38984	0.37043	0.35214	0.33490	0.31863	0.30328
7	0.35383	0.33320	0.31393	0.29592	0.27908	0.26333	0.24859
8	0.30503	0.28478	0.26604	0.24867	0.23257	0.21763	0.20376
9	0.26295	0.24340	0.22546	0.20897	0.19381	0.17986	0.16702
10	0.22668	0.20804	0.19106	0.17560	0.16151	0.14864	0.13690
11	0.19542	0.17781	0.16192	0.14757	0.13459	0.12285	0.11221
12	0.16846	0.15197	0.13722	0.12400	0.11216	0.10153	0.09198
13	0.14523	0.12989	0.11629	0.10421	0.09346	0.08391	0.07539
14	0.12520	0.11102	0.09855	0.08757	0.07789	0.06934	0.06180
15	0.10793	0.09489	0.08352	0.07359	0.06491	0.05731	0.05065
16	0.09304	0.08110	0.07078	0.06184	0.05409	0.04736	0.04152
17	0.08021	0.06932	0.05998	0.05196	0.04507	0.03914	0.03403
18	0.06914	0.05925	0.05083	0.04367	0.03756	0.03235	0.02789
19	0.05961	0.05064	0.04308	0.03670	0.03130	0.02673	0.02286
20	0.05139	0.04328	0.03651	0.03084	0.02608	0.02209	0.01874
21	0.04430	0.03699	0.03094	0.02591	0.02174	0.01826	0.01536
22	0.03819	0.03162	0.02622	0.02178	0.01811	0.01509	0.01259
23	0.03292	0.02702	0.02222	0.01830	0.01509	0.01247	0.01032
24	0.02838	0.02310	0.01883	0.01538	0.01258	0.01031	0.00846
25	0.02447	0.01974	0.01596	0.01292	0.01048	0.00852	0.00693
26	0.02109	0.01687	0.01352	0.01086	0.00874	0.00704	0.00568
27	0.01818	0.01442	0.01146	0.00912	0.00728	0.00582	0.00466
28	0.01567	0.01233	0.00971	0.00767	0.00607	0.00481	0.00382
29	0.01351	0.01053	0.00823	0.00644	0.00506	0.00397	0.00313
30	0.01165	0.00900	0.00697	0.00541	0.00421	0.00328	0.00257
35	0.00555	0.00411	0.00305	0.00227	0.00169	0.00127	0.00095
40	0.00264	0.00187	0.00133	0.00095	0.00068	0.00049	0.00035
45	0.00126	0.00085	0.00058	0.00040	0.00027	0.00019	0.00013
50	0.00060	0.00039	0.00025	0.00017	0.00011	0.00007	0.00005
60	0.00014	0.00008	0.00005	0.00003	0.00002	0.00001	0.00001
70	0.00003	0.00002	0.00001	0.00001	0.00000	0.00000	0.00000
80	0.00001	0.00000	0.00000	0.00000	0.00000	0.00000	0.00000

	Annual Interest Rates						
Year	23%	24%	25%	26%	27%	28%	29%
1	0.81301	0.80645	0.80000	0.79365	0.78740	0.78125	0.77519
2	0.66098	0.65036	0.64000	0.62988	0.62000	0.61035	0.60093
3	0.53738	0.52449	0.51200	0.49991	0.48819	0.47684	0.46583
4	0.43690	0.42297	0.40960	0.39675	0.38440	0.37253	0.36111
5	0.35520	0.34111	0.32768	0.31488	0.30268	0.29104	0.27993
6	0.28878	0.27509	0.26214	0.24991	0.23833	0.22737	0.21700
7	0.23478	0.22184	0.20972	0.19834	0.18766	0.17764	0.16822
8	0.19088	0.17891	0.16777	0.15741	0.14776	0.13878	0.13040
9	0.15519	0.14428	0.13422	0.12493	0.11635	0.10842	0.10109
10	0.12617	0.11635	0.10737	0.09915	0.09161	0.08470	0.07836
11	0.10258	0.09383	0.08590	0.07869	0.07214	0.06617	0.06075
12	0.08339	0.07567	0.06872	0.06245	0.05680	0.05170	0.04709
13	0.06780	0.06103	0.05498	0.04957	0.04473	0.04039	0.03650
14	0.05512	0.04921	0.04398	0.03934	0.03522	0.03155	0.02830
15	0.04481	0.03969	0.03518	0.03122	0.02773	0.02465	0.02194
16	0.03643	0.03201	0.02815	0.02478	0.02183	0.01926	0.01700
17	0.02962	0.02581	0.02252	0.01967	0.01719	0.01505	0.01318
18	0.02408	0.02082	0.01801	0.01561	0.01354	0.01175	0.01022
19	0.01958	0.01679	0.01441	0.01239	0.01066	0.00918	0.00792
20	0.01592	0.01354	0.01153	0.00983	0.00839	0.00717	0.00614
21	0.01294	0.01092	0.00922	0.00780	0.00661	0.00561	0.00476
22	0.01052	0.00880	0.00738	0.00619	0.00520	0.00438	0.00369
23	0.00855	0.00710	0.00590	0.00491	0.00410	0.00342	0.00286
24	0.00695	0.00573	0.00472	0.00390	0.00323	0.00267	0.00222
25	0.00565	0.00462	0.00378	0.00310	0.00254	0.00209	0.00172
26	0.00460	0.00372	0.00302	0.00246	0.00200	0.00163	0.00133
27	0.00374	0.00300	0.00242	0.00195	0.00158	0.00127	0.00103
28	0.00304	0.00242	0.00193	0.00155	0.00124	0.00100	0.00080
29	0.00247	0.00195	0.00155	0.00123	0.00098	0.00078	0.00062
30	0.00201	0.00158	0.00124	0.00097	0.00077	0.00061	0.00048
35	0.00071	0.00054	0.00041	0.00031	0.00023	0.00018	0.00013
40	0.00025	0.00018	0.00013	0.00010	0.00007	0.00005	0.00004
45	0.00009	0.00006	0.00004	0.00003	0.00002	0.00001	0.00001
50	0.00003	0.00002	0.00001	0.00001	0.00001	0.00000	0.00000

Year	Annual Interest Rates						
	30%	31%	32%	33%	34%	35%	36%
1	0.76923	0.76336	0.75758	0.75188	0.74627	0.74074	0.73529
2	0.59172	0.58272	0.57392	0.56532	0.55692	0.54870	0.54066
3	0.45517	0.44482	0.43479	0.42505	0.41561	0.40644	0.39754
4	0.35013	0.33956	0.32939	0.31959	0.31016	0.30107	0.29231
5	0.26933	0.25921	0.24953	0.24029	0.23146	0.22301	0.21493
6	0.20718	0.19787	0.18904	0.18067	0.17273	0.16520	0.15804
7	0.15937	0.15104	0.14321	0.13584	0.12890	0.12237	0.11621
8	0.12259	0.11530	0.10849	0.10214	0.09620	0.09064	0.08545
9	0.09430	0.08802	0.08219	0.07680	0.07179	0.06714	0.06283
10	0.07254	0.06719	0.06227	0.05774	0.05357	0.04974	0.04620
11	0.05580	0.05129	0.04717	0.04341	0.03998	0.03684	0.03397
12	0.04292	0.03915	0.03574	0.03264	0.02984	0.02729	0.02498
13	0.03302	0.02989	0.02707	0.02454	0.02227	0.02021	0.01837
14	0.02540	0.02281	0.02051	0.01845	0.01662	0.01497	0.01350
15	0.01954	0.01742	0.01554	0.01387	0.01240	0.01109	0.00993
16	0.01503	0.01329	0.01177	0.01043	0.00925	0.00822	0.00730
17	0.01156	0.01015	0.00892	0.00784	0.00691	0.00609	0.00537
18	0.00889	0.00775	0.00676	0.00590	0.00515	0.00451	0.00395
19	0.00684	0.00591	0.00512	0.00443	0.00385	0.00334	0.00290
20	0.00526	0.00451	0.00388	0.00333	0.00287	0.00247	0.00213
21	0.00405	0.00345	0.00294	0.00251	0.00214	0.00183	0.00157
22	0.00311	0.00263	0.00223	0.00188	0.00160	0.00136	0.00115
23	0.00239	0.00201	0.00169	0.00142	0.00119	0.00101	0.00085
24	0.00184	0.00153	0.00128	0.00107	0.00089	0.00074	0.00062
25	0.00142	0.00117	0.00097	0.00080	0.00066	0.00055	0.00046
26	0.00109	0.00089	0.00073	0.00060	0.00050	0.00041	0.00034
27	0.00084	0.00068	0.00056	0.00045	0.00037	0.00030	0.00025
28	0.00065	0.00052	0.00042	0.00034	0.00028	0.00022	0.00018
29	0.00050	0.00040	0.00032	0.00026	0.00021	0.00017	0.00013
30	0.00038	0.00030	0.00024	0.00019	0.00015	0.00012	0.00010
35	0.00010	0.00008	0.00006	0.00005	0.00004	0.00003	0.00002
40	0.00003	0.00002	0.00002	0.00001	0.00001	0.00001	0.00000
45	0.00001	0.00001	0.00000	0.00000	0.00000	0.00000	0.00000

			Annual Interest Rates				
Year	37%	38%	39%	40%	41%	42%	43%
1	0.72993	0.72464	0.71942	0.71429	0.70922	0.70423	0.69930
2	0.53279	0.52510	0.51757	0.51020	0.50299	0.49593	0.48902
3	0.38890	0.38051	0.37235	0.36443	0.35673	0.34925	0.34197
4	0.28387	0.27573	0.26788	0.26031	0.25300	0.24595	0.23914
5	0.20720	0.19980	0.19272	0.18593	0.17943	0.17320	0.16723
6	0.15124	0.14479	0.13865	0.13281	0.12726	0.12197	0.11695
7	0.11040	0.10492	0.09975	0.09486	0.09025	0.08590	0.08178
8	0.08058	0.07603	0.07176	0.06776	0.06401	0.06049	0.05719
9	0.05882	0.05509	0.05163	0.04840	0.04540	0.04260	0.03999
10	0.04293	0.03992	0.03714	0.03457	0.03220	0.03000	0.02797
11	0.03134	0.02893	0.02672	0.02469	0.02283	0.02113	0.01956
12	0.02287	0.02096	0.01922	0.01764	0.01619	0.01488	0.01368
13	0.01670	0.01519	0.01383	0.01260	0.01149	0.01048	0.00956
14	0.01219	0.01101	0.00995	0.00900	0.00815	0.00738	0.00669
15	0.00890	0.00798	0.00716	0.00643	0.00578	0.00520	0.00468
16	0.00649	0.00578	0.00515	0.00459	0.00410	0.00366	0.00327
17	0.00474	0.00419	0.00370	0.00328	0.00291	0.00258	0.00229
18	0.00346	0.00304	0.00267	0.00234	0.00206	0.00181	0.00160
19	0.00253	0.00220	0.00192	0.00167	0.00146	0.00128	0.00112
20	0.00184	0.00159	0.00138	0.00120	0.00104	0.00090	0.00078
21	0.00135	0.00115	0.00099	0.00085	0.00074	0.00063	0.00055
22	0.00098	0.00084	0.00071	0.00061	0.00052	0.00045	0.00038
23	0.00072	0.00061	0.00051	0.00044	0.00037	0.00031	0.00027
24	0.00052	0.00044	0.00037	0.00031	0.00026	0.00022	0.00019
25	0.00038	0.00032	0.00027	0.00022	0.00019	0.00016	0.00013
26	0.00028	0.00023	0.00019	0.00016	0.00013	0.00011	0.00009
27	0.00020	0.00017	0.00014	0.00011	0.00009	0.00008	0.00006
28	0.00015	0.00012	0.00010	0.00008	0.00007	0.00005	0.00004
29	0.00011	0.00009	0.00007	0.00006	0.00005	0.00004	0.00003
30	0.00008	0.00006	0.00005	0.00004	0.00003	0.00003	0.00002
35	0.00002	0.00001	0.00001	0.00001	0.00001	0.00000	0.00000

Year	Annual Interest Rates						
	44%	**45%**	**46%**	**47%**	**48%**	**49%**	**50%**
1	0.69444	0.68966	0.68493	0.68027	0.67568	0.67114	0.66667
2	0.48225	0.47562	0.46913	0.46277	0.45654	0.45043	0.44444
3	0.33490	0.32802	0.32132	0.31481	0.30847	0.30230	0.29630
4	0.23257	0.22622	0.22008	0.21416	0.20843	0.20289	0.19753
5	0.16151	0.15601	0.15074	0.14568	0.14083	0.13617	0.13169
6	0.11216	0.10759	0.10325	0.09911	0.09515	0.09139	0.08779
7	0.07789	0.07420	0.07072	0.06742	0.06429	0.06133	0.05853
8	0.05409	0.05117	0.04844	0.04586	0.04344	0.04116	0.03902
9	0.03756	0.03529	0.03318	0.03120	0.02935	0.02763	0.02601
10	0.02608	0.02434	0.02272	0.02122	0.01983	0.01854	0.01734
11	0.01811	0.01679	0.01556	0.01444	0.01340	0.01244	0.01156
12	0.01258	0.01158	0.01066	0.00982	0.00905	0.00835	0.00771
13	0.00874	0.00798	0.00730	0.00668	0.00612	0.00561	0.00514
14	0.00607	0.00551	0.00500	0.00455	0.00413	0.00376	0.00343
15	0.00421	0.00380	0.00343	0.00309	0.00279	0.00252	0.00228
16	0.00293	0.00262	0.00235	0.00210	0.00189	0.00169	0.00152
17	0.00203	0.00181	0.00161	0.00143	0.00128	0.00114	0.00101
18	0.00141	0.00125	0.00110	0.00097	0.00086	0.00076	0.00068
19	0.00098	0.00086	0.00075	0.00066	0.00058	0.00051	0.00045
20	0.00068	0.00059	0.00052	0.00045	0.00039	0.00034	0.00030
21	0.00047	0.00041	0.00035	0.00031	0.00027	0.00023	0.00020
22	0.00033	0.00028	0.00024	0.00021	0.00018	0.00015	0.00013
23	0.00023	0.00019	0.00017	0.00014	0.00012	0.00010	0.00009
24	0.00016	0.00013	0.00011	0.00010	0.00008	0.00007	0.00006
25	0.00011	0.00009	0.00008	0.00007	0.00006	0.00005	0.00004
26	0.00008	0.00006	0.00005	0.00004	0.00004	0.00003	0.00003
27	0.00005	0.00004	0.00004	0.00003	0.00003	0.00002	0.00002
28	0.00004	0.00003	0.00003	0.00002	0.00002	0.00001	0.00001
29	0.00003	0.00002	0.00002	0.00001	0.00001	0.00001	0.00001
30	0.00002	0.00001	0.00001	0.00001	0.00001	0.00001	0.00001

Appendix B

DISCOUNT FACTORS FOR ANNUAL COMPOUNDING AND MIDYEAR CASH FLOWS

			Annual Interest Rates				
Year	1%	2%	3%	4%	5%	6%	7%
1	0.99504	0.99015	0.98533	0.98058	0.97590	0.97129	0.96674
2	0.98519	0.97073	0.95663	0.94287	0.92943	0.91631	0.90349
3	0.97543	0.95170	0.92877	0.90660	0.88517	0.86444	0.84439
4	0.96577	0.93304	0.90172	0.87173	0.84302	0.81551	0.78915
5	0.95621	0.91474	0.87545	0.83820	0.80288	0.76935	0.73752
6	0.94674	0.89681	0.84995	0.80597	0.76464	0.72580	0.68927
7	0.93737	0.87922	0.82520	0.77497	0.72823	0.68472	0.64418
8	0.92809	0.86198	0.80116	0.74516	0.69355	0.64596	0.60203
9	0.91890	0.84508	0.77783	0.71650	0.66053	0.60940	0.56265
10	0.90980	0.82851	0.75517	0.68894	0.62907	0.57490	0.52584
11	0.90079	0.81227	0.73318	0.66245	0.59912	0.54236	0.49144
12	0.89188	0.79634	0.71182	0.63697	0.57059	0.51166	0.45929
13	0.88305	0.78072	0.69109	0.61247	0.54342	0.48270	0.42924
14	0.87430	0.76542	0.67096	0.58891	0.51754	0.45538	0.40116
15	0.86565	0.75041	0.65142	0.56626	0.49290	0.42960	0.37492
16	0.85707	0.73569	0.63245	0.54448	0.46942	0.40528	0.35039
17	0.84859	0.72127	0.61402	0.52354	0.44707	0.38234	0.32747
18	0.84019	0.70713	0.59614	0.50340	0.42578	0.36070	0.30604
19	0.83187	0.69326	0.57878	0.48404	0.40551	0.34028	0.28602
20	0.82363	0.67967	0.56192	0.46543	0.38620	0.32102	0.26731
21	0.81548	0.66634	0.54555	0.44752	0.36781	0.30285	0.24982
22	0.80740	0.65328	0.52966	0.43031	0.35029	0.28571	0.23348
23	0.79941	0.64047	0.51424	0.41376	0.33361	0.26954	0.21821
24	0.79149	0.62791	0.49926	0.39785	0.31773	0.25428	0.20393
25	0.78366	0.61560	0.48472	0.38255	0.30260	0.23989	0.19059
26	0.77590	0.60353	0.47060	0.36783	0.28819	0.22631	0.17812
27	0.76822	0.59169	0.45689	0.35369	0.27446	0.21350	0.16647
28	0.76061	0.56009	0.44358	0.34008	0.26139	0.20141	0.15558
29	0.75308	0.56872	0.43066	0.32700	0.24895	0.19001	0.14540
30	0.74562	0.55756	0.41812	0.31442	0.23709	0.17926	0.13589
31	0.73624	0.54663	0.40594	0.30233	0.22580	0.16911	0.12700
32	0.73093	0.53591	0.39412	0.29070	0.21505	0.15954	0.11869
33	0.72369	0.52541	0.38264	0.27952	0.20481	0.15051	0.11092
34	0.71653	0.51510	0.37150	0.26877	0.19506	0.14199	0.10367
35	0.70944	0.50500	0.36068	0.25843	0.18577	0.13395	0.09689
36	0.70241	0.49510	0.35017	0.24849	0.17692	0.12637	0.09055
37	0.69546	0.48539	0.33997	0.23894	0.16850	0.11922	0.08462
38	0.68857	0.47588	0.33007	0.22975	0.16047	0.11247	0.07909
39	0.68175	0.46654	0.32046	0.22091	0.15283	0.10610	0.07391
40	0.67500	0.45740	0.31112	0.21241	0.14555	0.10010	0.06908

			Annual Interest Rates				
Year	1%	2%	3%	4%	5%	6%	7%
41	0.66832	0.44843	0.30206	0.20424	0.13862	0.09443	0.06456
42	0.66170	0.43964	0.29326	0.19639	0.13202	0.08909	0.06034
43	0.65515	0.43102	0.28472	0.18884	0.12573	0.08404	0.05639
44	0.64867	0.42256	0.27643	0.18157	0.11975	0.07929	0.05270
45	0.64224	0.41428	0.26838	0.17459	0.11405	0.07480	0.04925
46	0.63588	0.40616	0.26056	0.16787	0.10861	0.07056	0.04603
47	0.62959	0.39819	0.25297	0.16142	0.10344	0.06657	0.04302
48	0.62335	0.39038	0.24560	0.15521	0.09852	0.06280	0.04020
49	0.61718	0.38273	0.23845	0.14924	0.09383	0.05925	0.03757
50	0.61107	0.37522	0.23150	0.14350	0.08936	0.05589	0.03512
60	0.55320	0.30782	0.17226	0.09694	0.05486	0.03121	0.01765
70	0.50080	0.25252	0.12818	0.06549	0.03368	0.01743	0.00907
80	0.45337	0.20715	0.09538	0.04424	0.02068	0.00973	0.00461
90	0.41043	0.16994	0.07097	0.02989	0.01269	0.00543	0.00235
100	0.37156	0.13941	0.05281	0.02019	0.00779	0.00303	0.00119

			Annual Interest Rates				
Year	8%	9%	10%	15%	20%	25%	30%
1	0.96225	0.95783	0.95346	0.93250	0.91287	0.89443	0.87706
2	0.89097	0.87874	0.86678	0.81087	0.76073	0.71554	0.67466
3	0.82497	0.80618	0.78799	0.70511	0.63394	0.57243	0.51897
4	0.76387	0.73962	0.71635	0.61314	0.52828	0.45795	0.39921
5	0.70728	0.67855	0.65123	0.53316	0.44023	0.36636	0.30708
6	0.65489	0.62252	0.59203	0.46362	0.36686	0.29309	0.23622
7	0.60638	0.57112	0.53820	0.40315	0.30572	0.23447	0.18171
8	0.56146	0.52396	0.48928	0.35056	0.25477	0.18758	0.13977
9	0.51987	0.48070	0.44480	0.30484	0.21230	0.15006	0.10752
10	0.48136	0.44101	0.40436	0.26508	0.17692	0.12005	0.08271
11	0.44571	0.40460	0.36760	0.23050	0.14743	0.09604	0.06362
12	0.41269	0.37119	0.33418	0.20044	0.12286	0.07683	0.04894
13	0.38212	0.34054	0.30380	0.17429	0.10238	0.06146	0.03765
14	0.35382	0.31242	0.27618	0.15156	0.08532	0.04917	0.02896
15	0.32761	0.28663	0.25108	0.13179	0.07110	0.03934	0.02228
16	0.30334	0.26296	0.22825	0.11460	0.05925	0.03147	0.01713
17	0.28087	0.24125	0.20750	0.09965	0.04938	0.02518	0.01318
18	0.26007	0.22133	0.18864	0.08665	0.04115	0.02014	0.01014
19	0.24080	0.20305	0.17149	0.07535	0.03429	0.01611	0.00780
20	0.22297	0.18629	0.15590	0.06552	0.02857	0.01289	0.00600

Year	8%	9%	10%	15%	20%	25%	30%
				Annual Interest Rates			
21	0.20645	0.17091	0.14173	0.05698	0.02381	0.01031	0.00461
22	0.19116	0.15679	0.12884	0.04954	0.01984	0.00825	0.00355
23	0.17700	0.14385	0.11713	0.04308	0.01654	0.00660	0.00273
24	0.16389	0.13197	0.10648	0.03746	0.01378	0.00528	0.00210
25	0.15175	0.12107	0.09680	0.03258	0.01148	0.00422	0.00162
26	0.14051	0.11108	0.08800	0.02833	0.00957	0.00338	0.00124
27	0.13010	0.10191	0.08000	0.02463	0.00797	0.00270	0.00096
28	0.12046	0.09349	0.07273	0.02142	0.00665	0.00216	0.00074
29	0.11154	0.08577	0.06612	0.01863	0.00554	0.00173	0.00057
30	0.10328	0.07869	0.06011	0.01620	0.00461	0.00138	0.00044
31	0.09563	0.07219	0.05464	0.01408	0.00385	0.00111	0.00033
32	0.08854	0.06623	0.04967	0.01225	0.00320	0.00089	0.00026
33	0.08198	0.06076	0.04516	0.01065	0.00267	0.00071	0.00020
34	0.07591	0.05575	0.04105	0.00926	0.00223	0.00057	0.00015
35	0.07029	0.05114	0.03732	0.00805	0.00185	0.00045	0.00012
36	0.06508	0.04692	0.03393	0.00700	0.00155	0.00036	0.00009
37	0.06026	0.04305	0.03084	0.00609	0.00129	0.00029	0.00007
38	0.05580	0.03949	0.02804	0.00529	0.00107	0.00023	0.00005
39	0.05166	0.03623	0.02549	0.00460	0.00089	0.00019	0.00004
40	0.04784	0.03324	0.02317	0.00400	0.00075	0.00015	0.00003
41	0.04429	0.03049	0.02107	0.00348	0.00062	0.00012	0.00002
42	0.04101	0.02798	0.01915	0.00303	0.00052	0.00010	0.00002
43	0.03797	0.02567	0.01741	0.00263	0.00043	0.00008	0.00001
44	0.03516	0.02355	0.01583	0.00229	0.00036	0.00006	0.00001
45	0.03256	0.02160	0.01439	0.00199	0.00030	0.00005	0.00001
46	0.03015	0.01982	0.01308	0.00173	0.00025	0.00004	0.00001
47	0.02791	0.01818	0.01189	0.00151	0.00021	0.00003	0.00001
48	0.02584	0.01668	0.01081	0.00131	0.00017	0.00002	0.00000
49	0.02393	0.01530	0.00983	0.00114	0.00014	0.00002	0.00000
50	0.02216	0.01404	0.00893	0.00099	0.00012	0.00002	0.00000
60	0.01026	0.00593	0.00344	0.00024	0.00002	0.00000	0.00000
70	0.00475	0.00251	0.00133	0.00006	0.00000	0.00000	0.00000
80	0.00220	0.00106	0.00051	0.00001	0.00000	0.00000	0.00000
90	0.00102	0.00045	0.00020	0.00000	0.00000	0.00000	0.00000
100	0.00047	0.00019	0.00008	0.00000	0.00000	0.00000	0.00000

			Annual Interest Rates				
Year	35%	40%	45%	50%	55%	60%	65%
1	0.86066	0.84515	0.83045	0.81650	0.80322	0.79057	0.77850
2	0.63753	0.60368	0.57273	0.54433	0.51821	0.49411	0.47182
3	0.47224	0.43120	0.39498	0.36289	0.33433	0.30882	0.28595
4	0.34981	0.30800	0.27240	0.24192	0.21569	0.19301	0.17330
5	0.25912	0.22000	0.18786	0.16128	0.13916	0.12063	0.10503
6	0.19194	0.15714	0.12956	0.10752	0.08978	0.07539	0.06366
7	0.14218	0.11225	0.08935	0.07168	0.05792	0.04712	0.03858
8	0.10532	0.08018	0.06162	0.04779	0.03737	0.02945	0.02338
9	0.07801	0.05727	0.04250	0.03186	0.02411	0.01841	0.01417
10	0.05779	0.04091	0.02931	0.02124	0.01555	0.01150	0.00859
11	0.04281	0.02922	0.02021	0.01416	0.01004	0.00719	0.00520
12	0.03171	0.02087	0.01394	0.00944	0.00647	0.00449	0.00315
13	0.02349	0.01491	0.00961	0.00629	0.00418	0.00281	0.00191
14	0.01740	0.01065	0.00663	0.00420	0.00269	0.00176	0.00116
15	0.01289	0.00761	0.00457	0.00280	0.00174	0.00110	0.00070
16	0.00955	0.00543	0.00315	0.00186	0.00112	0.00069	0.00043
17	0.00707	0.00388	0.00217	0.00124	0.00072	0.00043	0.00026
18	0.00524	0.00277	0.00150	0.00083	0.00047	0.00027	0.00016
19	0.00388	0.00198	0.00103	0.00055	0.00030	0.00017	0.00009
20	0.00287	0.00141	0.00071	0.00037	0.00019	0.00010	0.00006
21	0.00213	0.00101	0.00049	0.00025	0.00013	0.00007	0.00003
22	0.00158	0.00072	0.00034	0.00016	0.00008	0.00004	0.00002
23	0.00117	0.00052	0.00023	0.00011	0.00005	0.00003	0.00001
24	0.00087	0.00037	0.00016	0.00007	0.00003	0.00002	0.00001
25	0.00064	0.00026	0.00011	0.00005	0.00002	0.00001	0.00000
26	0.00047	0.00019	0.00008	0.00003	0.00001	0.00001	0.00000
27	0.00035	0.00013	0.00005	0.00002	0.00001	0.00000	0.00000
28	0.00026	0.00010	0.00004	0.00001	0.00001	0.00000	0.00000
29	0.00019	0.00007	0.00003	0.00001	0.00000	0.00000	0.00000
30	0.00014	0.00005	0.00002	0.00001	0.00000	0.00000	0.00000
31	0.00011	0.00003	0.00001	0.00000	0.00000	0.00000	0.00000
32	0.00008	0.00002	0.00001	0.00000	0.00000	0.00000	0.00000
33	0.00006	0.00002	0.00001	0.00000	0.00000	0.00000	0.00000
34	0.00004	0.00001	0.00000	0.00000	0.00000	0.00000	0.00000
35	0.00003	0.00001	0.00000	0.00000	0.00000	0.00000	0.00000
36	0.00002	0.00001	0.00000	0.00000	0.00000	0.00000	0.00000
37	0.00002	0.00000	0.00000	0.00000	0.00000	0.00000	0.00000
38	0.00001	0.00000	0.00000	0.00000	0.00000	0.00000	0.00000
39	0.00001	0.00000	0.00000	0.00000	0.00000	0.00000	0.00000
40	0.00001	0.00000	0.00000	0.00000	0.00000	0.00000	0.00000

			Annual Interest Rates				
Year	35%	40%	45%	50%	55%	60%	65%
41	0.00001	0.00000	0.00000	0.00000	0.00000	0.00000	0.00000
42	0.00000	0.00000	0.00000	0.00000	0.00000	0.00000	0.00000
43	0.00000	0.00000	0.00000	0.00000	0.00000	0.00000	0.00000
44	0.00000	0.00000	0.00000	0.00000	0.00000	0.00000	0.00000
45	0.00000	0.00000	0.00000	0.00000	0.00000	0.00000	0.00000
46	0.00000	0.00000	0.00000	0.00000	0.00000	0.00000	0.00000
47	0.00000	0.00000	0.00000	0.00000	0.00000	0.00000	0.00000
48	0.00000	0.00000	0.00000	0.00000	0.00000	0.00000	0.00000
49	0.00000	0.00000	0.00000	0.00000	0.00000	0.00000	0.00000
50	0.00000	0.00000	0.00000	0.00000	0.00000	0.00000	0.00000
60	0.00000	0.00000	0.00000	0.00000	0.00000	0.00000	0.00000
70	0.00000	0.00000	0.00000	0.00000	0.00000	0.00000	0.00000
80	0.00000	0.00000	0.00000	0.00000	0.00000	0.00000	0.00000
90	0.00000	0.00000	0.00000	0.00000	0.00000	0.00000	0.00000
100	0.00000	0.00000	0.00000	0.00000	0.00000	0.00000	0.00000

			Annual Interest Rates				
Year	70%	75%	80%	85%	90%	95%	100%
1	0.76697	0.75593	0.74536	0.73521	0.72548	0.71611	0.70711
2	0.45116	0.43196	0.41409	0.39741	0.38183	0.36724	0.35355
3	0.26539	0.24683	0.23005	0.21482	0.20096	0.18833	0.17678
4	0.15611	0.14105	0.12780	0.11612	0.10577	0.09658	0.08839
5	0.09183	0.08060	0.07100	0.06277	0.05567	0.04953	0.04419
6	0.05402	0.04606	0.03945	0.03393	0.02930	0.02540	0.02210
7	0.03177	0.02632	0.02191	0.01834	0.01542	0.01302	0.01105
8	0.01869	0.01504	0.01217	0.00991	0.00812	0.00668	0.00552
9	0.01099	0.00859	0.00676	0.00536	0.00427	0.00343	0.00276
10	0.00647	0.00491	0.00376	0.00290	0.00225	0.00176	0.00138
11	0.00380	0.00281	0.00209	0.00157	0.00118	0.00090	0.00069
12	0.00224	0.00160	0.00116	0.00085	0.00062	0.00046	0.00035
13	0.00132	0.00092	0.00064	0.00046	0.00033	0.00024	0.00017
14	0.00077	0.00052	0.00036	0.00025	0.00017	0.00012	0.00009
15	0.00046	0.00030	0.00020	0.00013	0.00009	0.00006	0.00004
16	0.00027	0.00017	0.00011	0.00007	0.00005	0.00003	0.00002
17	0.00016	0.00010	0.00006	0.00004	0.00003	0.00002	0.00001
18	0.00009	0.00006	0.00003	0.00002	0.00001	0.00001	0.00001
19	0.00005	0.00003	0.00002	0.00001	0.00001	0.00000	0.00000
20	0.00003	0.00002	0.00001	0.00001	0.00000	0.00000	0.00000

			Annual Interest Rates				
Year	**35%**	**40%**	**45%**	**50%**	**55%**	**60%**	**65%**
21	0.00002	0.00001	0.00001	0.00000	0.00000	0.00000	0.00000
22	0.00001	0.00001	0.00000	0.00000	0.00000	0.00000	0.00000
23	0.00001	0.00000	0.00000	0.00000	0.00000	0.00000	0.00000
24	0.00000	0.00000	0.00000	0.00000	0.00000	0.00000	0.00000
25	0.00000	0.00000	0.00000	0.00000	0.00000	0.00000	0.00000
26	0.00000	0.00000	0.00000	0.00000	0.00000	0.00000	0.00000
27	0.00000	0.00000	0.00000	0.00000	0.00000	0.00000	0.00000
28	0.00000	0.00000	0.00000	0.00000	0.00000	0.00000	0.00000
29	0.00000	0.00000	0.00000	0.00000	0.00000	0.00000	0.00000
30	0.00000	0.00000	0.00000	0.00000	0.00000	0.00000	0.00000
31	0.00000	0.00000	0.00000	0.00000	0.00000	0.00000	0.00000
32	0.00000	0.00000	0.00000	0.00000	0.00000	0.00000	0.00000
33	0.00000	0.00000	0.00000	0.00000	0.00000	0.00000	0.00000
34	0.00000	0.00000	0.00000	0.00000	0.00000	0.00000	0.00000
35	0.00000	0.00000	0.00000	0.00000	0.00000	0.00000	0.00000
36	0.00000	0.00000	0.00000	0.00000	0.00000	0.00000	0.00000
37	0.00000	0.00000	0.00000	0.00000	0.00000	0.00000	0.00000
38	0.00000	0.00000	0.00000	0.00000	0.00000	0.00000	0.00000
39	0.00000	0.00000	0.00000	0.00000	0.00000	0.00000	0.00000
40	0.00000	0.00000	0.00000	0.00000	0.00000	0.00000	0.00000
41	0.00000	0.00000	0.00000	0.00000	0.00000	0.00000	0.00000
42	0.00000	0.00000	0.00000	0.00000	0.00000	0.00000	0.00000
43	0.00000	0.00000	0.00000	0.00000	0.00000	0.00000	0.00000
44	0.00000	0.00000	0.00000	0.00000	0.00000	0.00000	0.00000
45	0.00000	0.00000	0.00000	0.00000	0.00000	0.00000	0.00000
46	0.00000	0.00000	0.00000	0.00000	0.00000	0.00000	0.00000
47	0.00000	0.00000	0.00000	0.00000	0.00000	0.00000	0.00000
48	0.00000	0.00000	0.00000	0.00000	0.00000	0.00000	0.00000
49	0.00000	0.00000	0.00000	0.00000	0.00000	0.00000	0.00000
50	0.00000	0.00000	0.00000	0.00000	0.00000	0.00000	0.00000
60	0.00000	0.00000	0.00000	0.00000	0.00000	0.00000	0.00000
70	0.00000	0.00000	0.00000	0.00000	0.00000	0.00000	0.00000
80	0.00000	0.00000	0.00000	0.00000	0.00000	0.00000	0.00000
90	0.00000	0.00000	0.00000	0.00000	0.00000	0.00000	0.00000
100	0.00000	0.00000	0.00000	0.00000	0.00000	0.00000	0.00000

Appendix C

DISCOUNT FACTORS FOR CONTINUOUS COMPOUNDING AND YEAR-END CASH FLOWS

Year	Continuous Interest Rates						
	2%	3%	4%	5%	6%	7%	8%
1	0.98020	0.97045	0.96079	0.95123	0.94176	0.93239	0.92312
2	0.96079	0.94176	0.92312	0.90484	0.88692	0.86936	0.85214
3	0.94176	0.91393	0.88692	0.86071	0.83527	0.81058	0.78663
4	0.92312	0.88692	0.85214	0.81873	0.78663	0.75578	0.72615
5	0.90484	0.86071	0.81873	0.77880	0.74082	0.70469	0.67032
6	0.88692	0.83527	0.78663	0.74082	0.69768	0.65705	0.61878
7	0.86936	0.81058	0.75578	0.70469	0.65705	0.61263	0.57121
8	0.85214	0.78663	0.72615	0.67032	0.61878	0.57121	0.52729
9	0.83527	0.76338	0.69768	0.63763	0.58275	0.53259	0.48675
10	0.81873	0.74082	0.67032	0.60653	0.54881	0.49659	0.44933
11	0.80252	0.71892	0.64404	0.57695	0.51685	0.46301	0.41478
12	0.78663	0.69768	0.61878	0.54881	0.48675	0.43171	0.38289
13	0.77105	0.67706	0.59452	0.52205	0.45841	0.40252	0.35345
14	0.75578	0.65705	0.57121	0.49659	0.43171	0.37531	0.32628
15	0.74082	0.63763	0.54881	0.47237	0.40657	0.34994	0.30119
16	0.72615	0.61878	0.52729	0.44933	0.38289	0.32628	0.27804
17	0.71177	0.60050	0.50662	0.42741	0.36059	0.30422	0.25666
18	0.69768	0.58275	0.48675	0.40657	0.33960	0.28365	0.23693
19	0.68386	0.56553	0.46767	0.38674	0.31982	0.26448	0.21871
20	0.67032	0.54881	0.44933	0.36788	0.30119	0.24660	0.20190
21	0.65705	0.53259	0.43171	0.34994	0.28365	0.22993	0.18637
22	0.64404	0.51685	0.41478	0.33287	0.26714	0.21438	0.17204
23	0.63128	0.50158	0.39852	0.31664	0.25158	0.19989	0.15882
24	0.61878	0.48675	0.38289	0.30119	0.23693	0.18637	0.14661
25	0.60653	0.47237	0.36788	0.28650	0.22313	0.17377	0.13534
26	0.59452	0.45841	0.35345	0.27253	0.21014	0.16203	0.12493
27	0.58275	0.44486	0.33960	0.25924	0.19790	0.15107	0.11533
28	0.57121	0.43171	0.32628	0.24660	0.18637	0.14086	0.10646
29	0.55990	0.41895	0.31349	0.23457	0.17552	0.13134	0.09827
30	0.54881	0.40657	0.30119	0.22313	0.16530	0.12246	0.09072
35	0.49659	0.34994	0.24660	0.17377	0.12246	0.08629	0.06081
40	0.44933	0.30119	0.20190	0.13534	0.09072	0.06081	0.04076
45	0.40657	0.25924	0.16530	0.10540	0.06721	0.04285	0.02732
50	0.36788	0.22313	0.13534	0.08208	0.04979	0.03020	0.01832
60	0.30119	0.16530	0.09072	0.04979	0.02732	0.01500	0.00823
70	0.24660	0.12246	0.06081	0.03020	0.01500	0.00745	0.00370
80	0.20190	0.09072	0.04076	0.01832	0.00823	0.00370	0.00166
90	0.16530	0.06721	0.02732	0.01111	0.00452	0.00184	0.00075
100	0.13534	0.04979	0.01832	0.00674	0.00248	0.00091	0.00034

			Continuous Interest Rates				
Year	**9%**	**10%**	**11%**	**12%**	**13%**	**14%**	**15%**
1	0.91393	0.90484	0.89583	0.88692	0.87810	0.86936	0.86071
2	0.83527	0.81873	0.80252	0.78663	0.77105	0.75578	0.74082
3	0.76338	0.74082	0.71892	0.69768	0.67706	0.65705	0.63763
4	0.69768	0.67032	0.64404	0.61878	0.59452	0.57121	0.54881
5	0.63763	0.60653	0.57695	0.54881	0.52205	0.49659	0.47237
6	0.58275	0.54881	0.51685	0.48675	0.45841	0.43171	0.40657
7	0.53259	0.49659	0.46301	0.43171	0.40252	0.37531	0.34994
8	0.48675	0.44933	0.41478	0.38289	0.35345	0.32628	0.30119
9	0.44486	0.40657	0.37158	0.33960	0.31037	0.28365	0.25924
10	0.40657	0.36788	0.33287	0.30119	0.27253	0.24660	0.22313
11	0.37158	0.33287	0.29820	0.26714	0.23931	0.21438	0.19205
12	0.33960	0.30119	0.26714	0.23693	0.21014	0.18637	0.16530
13	0.31037	0.27253	0.23931	0.21014	0.18452	0.16203	0.14227
14	0.28365	0.24660	0.21438	0.18637	0.16203	0.14086	0.12246
15	0.25924	0.22313	0.19205	0.16530	0.14227	0.12246	0.10540
16	0.23693	0.20190	0.17204	0.14661	0.12493	0.10646	0.09072
17	0.21654	0.18268	0.15412	0.13003	0.10970	0.09255	0.07808
18	0.19790	0.16530	0.13807	0.11533	0.09633	0.08046	0.06721
19	0.18087	0.14957	0.12369	0.10228	0.08458	0.06995	0.05784
20	0.16530	0.13534	0.11080	0.09072	0.07427	0.06081	0.04979
21	0.15107	0.12246	0.09926	0.08046	0.06522	0.05287	0.04285
22	0.13807	0.11080	0.08892	0.07136	0.05727	0.04596	0.03688
23	0.12619	0.10026	0.07966	0.06329	0.05029	0.03996	0.03175
24	0.11533	0.09072	0.07136	0.05613	0.04416	0.03474	0.02732
25	0.10540	0.08208	0.06393	0.04979	0.03877	0.03020	0.02352
26	0.09633	0.07427	0.05727	0.04416	0.03405	0.02625	0.02024
27	0.08804	0.06721	0.05130	0.03916	0.02990	0.02282	0.01742
28	0.08046	0.06081	0.04596	0.03474	0.02625	0.01984	0.01500
29	0.07353	0.05502	0.04117	0.03081	0.02305	0.01725	0.01291
30	0.06721	0.04979	0.03688	0.02732	0.02024	0.01500	0.01111
35	0.04285	0.03020	0.02128	0.01500	0.01057	0.00745	0.00525
40	0.02732	0.01832	0.01228	0.00823	0.00552	0.00370	0.00248
45	0.01742	0.01111	0.00708	0.00452	0.00288	0.00184	0.00117
50	0.01111	0.00674	0.00409	0.00248	0.00150	0.00091	0.00055
60	0.00452	0.00248	0.00136	0.00075	0.00041	0.00022	0.00012
70	0.00184	0.00091	0.00045	0.00022	0.00011	0.00006	0.00003
80	0.00075	0.00034	0.00015	0.00007	0.00003	0.00001	0.00001
90	0.00030	0.00012	0.00005	0.00002	0.00001	0.00000	0.00000
100	0.00012	0.00005	0.00002	0.00001	0.00000	0.00000	0.00000

Year			Continuous Interest Rates				
	16%	17%	18%	19%	20%	21%	22%
1	0.85214	0.84366	0.83527	0.82696	0.81873	0.81058	0.80252
2	0.72615	0.71177	0.69768	0.68386	0.67032	0.65705	0.64404
3	0.61878	0.60050	0.58275	0.56553	0.54881	0.53259	0.51685
4	0.52729	0.50662	0.48675	0.46767	0.44933	0.43171	0.41478
5	0.44933	0.42741	0.40657	0.38674	0.36788	0.34994	0.33287
6	0.38289	0.36059	0.33960	0.31982	0.30119	0.28365	0.26714
7	0.32628	0.30422	0.28365	0.26448	0.24660	0.22993	0.21438
8	0.27804	0.25666	0.23693	0.21871	0.20190	0.18637	0.17204
9	0.23693	0.21654	0.19790	0.18087	0.16530	0.15107	0.13807
10	0.20190	0.18268	0.16530	0.14957	0.13534	0.12246	0.11080
11	0.17204	0.15412	0.13807	0.12369	0.11080	0.09926	0.08892
12	0.14661	0.13003	0.11533	0.10228	0.09072	0.08046	0.07136
13	0.12493	0.10970	0.09633	0.08458	0.07427	0.06522	0.05727
14	0.10646	0.09255	0.08046	0.06995	0.06081	0.05287	0.04596
15	0.09072	0.07808	0.06721	0.05784	0.04979	0.04285	0.03688
16	0.07730	0.06587	0.05613	0.04783	0.04076	0.03474	0.02960
17	0.06587	0.05558	0.04689	0.03956	0.03337	0.02816	0.02375
18	0.05613	0.04689	0.03916	0.03271	0.02732	0.02282	0.01906
19	0.04783	0.03956	0.03271	0.02705	0.02237	0.01850	0.01530
20	0.04076	0.03337	0.02732	0.02237	0.01832	0.01500	0.01228
21	0.03474	0.02816	0.02282	0.01850	0.01500	0.01216	0.00985
22	0.02960	0.02375	0.01906	0.01530	0.01228	0.00985	0.00791
23	0.02522	0.02004	0.01592	0.01265	0.01005	0.00799	0.00635
24	0.02149	0.01691	0.01330	0.01046	0.00823	0.00647	0.00509
25	0.01832	0.01426	0.01111	0.00865	0.00674	0.00525	0.00409
26	0.01561	0.01203	0.00928	0.00715	0.00552	0.00425	0.00328
27	0.01330	0.01015	0.00775	0.00592	0.00452	0.00345	0.00263
28	0.01133	0.00857	0.00647	0.00489	0.00370	0.00279	0.00211
29	0.00966	0.00723	0.00541	0.00405	0.00303	0.00227	0.00170
30	0.00823	0.00610	0.00452	0.00335	0.00248	0.00184	0.00136
35	0.00370	0.00261	0.00184	0.00129	0.00091	0.00064	0.00045
40	0.00166	0.00111	0.00075	0.00050	0.00034	0.00022	0.00015
45	0.00075	0.00048	0.00030	0.00019	0.00012	0.00008	0.00005
50	0.00034	0.00020	0.00012	0.00007	0.00005	0.00003	0.00002
60	0.00007	0.00004	0.00002	0.00001	0.00001	0.00000	0.00000
70	0.00001	0.00001	0.00000	0.00000	0.00000	0.00000	0.00000

	Continuous Interest Rates						
Year	23%	24%	25%	26%	27%	28%	29%
1	0.79453	0.78663	0.77880	0.77105	0.76338	0.75578	0.74826
2	0.63128	0.61878	0.60653	0.59452	0.58275	0.57121	0.55990
3	0.50158	0.48675	0.47237	0.45841	0.44486	0.43171	0.41895
4	0.39852	0.38289	0.36788	0.35345	0.33960	0.32628	0.31349
5	0.31664	0.30119	0.28650	0.27253	0.25924	0.24660	0.23457
6	0.25158	0.23693	0.22313	0.21014	0.19790	0.18637	0.17552
7	0.19989	0.18637	0.17377	0.16203	0.15107	0.14086	0.13134
8	0.15882	0.14661	0.13534	0.12493	0.11533	0.10646	0.09827
9	0.12619	0.11533	0.10540	0.09633	0.08804	0.08046	0.07353
10	0.10026	0.09072	0.08208	0.07427	0.06721	0.06081	0.05502
11	0.07966	0.07136	0.06393	0.05727	0.05130	0.04596	0.04117
12	0.06329	0.05613	0.04979	0.04416	0.03916	0.03474	0.03081
13	0.05029	0.04416	0.03877	0.03405	0.02990	0.02625	0.02305
14	0.03996	0.03474	0.03020	0.02625	0.02282	0.01984	0.01725
15	0.03175	0.02732	0.02352	0.02024	0.01742	0.01500	0.01291
16	0.02522	0.02149	0.01832	0.01561	0.01330	0.01133	0.00966
17	0.02004	0.01691	0.01426	0.01203	0.01015	0.00857	0.00723
18	0.01592	0.01330	0.01111	0.00928	0.00775	0.00647	0.00541
19	0.01265	0.01046	0.00865	0.00715	0.00592	0.00489	0.00405
20	0.01005	0.00823	0.00674	0.00552	0.00452	0.00370	0.00303
21	0.00799	0.00647	0.00525	0.00425	0.00345	0.00279	0.00227
22	0.00635	0.00509	0.00409	0.00328	0.00263	0.00211	0.00170
23	0.00504	0.00401	0.00318	0.00253	0.00201	0.00160	0.00127
24	0.00401	0.00315	0.00248	0.00195	0.00153	0.00121	0.00095
25	0.00318	0.00248	0.00193	0.00150	0.00117	0.00091	0.00071
26	0.00253	0.00195	0.00150	0.00116	0.00089	0.00069	0.00053
27	0.00201	0.00153	0.00117	0.00089	0.00068	0.00052	0.00040
28	0.00160	0.00121	0.00091	0.00069	0.00052	0.00039	0.00030
29	0.00127	0.00095	0.00071	0.00053	0.00040	0.00030	0.00022
30	0.00101	0.00075	0.00055	0.00041	0.00030	0.00022	0.00017
35	0.00032	0.00022	0.00016	0.00011	0.00008	0.00006	0.00004
40	0.00010	0.00007	0.00005	0.00003	0.00002	0.00001	0.00001
45	0.00003	0.00002	0.00001	0.00001	0.00001	0.00000	0.00000
50	0.00001	0.00001	0.00000	0.00000	0.00000	0.00000	0.00000

			Continuous Interest Rates				
Year	**30%**	**31%**	**32%**	**33%**	**34%**	**35%**	**36%**
1	0.74082	0.73345	0.72615	0.71892	0.71177	0.70469	0.69768
2	0.54881	0.53794	0.52729	0.51685	0.50662	0.49659	0.48675
3	0.40657	0.39455	0.38289	0.37158	0.36059	0.34994	0.33960
4	0.30119	0.28938	0.27804	0.26714	0.25666	0.24660	0.23693
5	0.22313	0.21225	0.20190	0.19205	0.18268	0.17377	0.16530
6	0.16530	0.15567	0.14661	0.13807	0.13003	0.12246	0.11533
7	0.12246	0.11418	0.10646	0.09926	0.09255	0.08629	0.08046
8	0.09072	0.08374	0.07730	0.07136	0.06587	0.06081	0.05613
9	0.06721	0.06142	0.05613	0.05130	0.04689	0.04285	0.03916
10	0.04979	0.04505	0.04076	0.03688	0.03337	0.03020	0.02732
11	0.03688	0.03304	0.02960	0.02652	0.02375	0.02128	0.01906
12	0.02732	0.02423	0.02149	0.01906	0.01691	0.01500	0.01330
13	0.02024	0.01777	0.01561	0.01370	0.01203	0.01057	0.00928
14	0.01500	0.01304	0.01133	0.00985	0.00857	0.00745	0.00647
15	0.01111	0.00956	0.00823	0.00708	0.00610	0.00525	0.00452
16	0.00823	0.00701	0.00598	0.00509	0.00434	0.00370	0.00315
17	0.00610	0.00514	0.00434	0.00366	0.00309	0.00261	0.00220
18	0.00452	0.00377	0.00315	0.00263	0.00220	0.00184	0.00153
19	0.00335	0.00277	0.00229	0.00189	0.00156	0.00129	0.00107
20	0.00248	0.00203	0.00166	0.00136	0.00111	0.00091	0.00075
21	0.00184	0.00149	0.00121	0.00098	0.00079	0.00064	0.00052
22	0.00136	0.00109	0.00088	0.00070	0.00056	0.00045	0.00036
23	0.00101	0.00080	0.00064	0.00051	0.00040	0.00032	0.00025
24	0.00075	0.00059	0.00046	0.00036	0.00029	0.00022	0.00018
25	0.00055	0.00043	0.00034	0.00026	0.00020	0.00016	0.00012
26	0.00041	0.00032	0.00024	0.00019	0.00014	0.00011	0.00009
27	0.00030	0.00023	0.00018	0.00014	0.00010	0.00008	0.00006
28	0.00022	0.00017	0.00013	0.00010	0.00007	0.00006	0.00004
29	0.00017	0.00012	0.00009	0.00007	0.00005	0.00004	0.00003
30	0.00012	0.00009	0.00007	0.00005	0.00004	0.00003	0.00002
35	0.00003	0.00002	0.00001	0.00001	0.00001	0.00000	0.00000
40	0.00001	0.00000	0.00000	0.00000	0.00000	0.00000	0.00000

| | **Continuous Interest Rates** | | | | | | |
Year	37%	38%	39%	40%	41%	42%	43%
1	0.69073	0.68386	0.67706	0.67032	0.66365	0.65705	0.65051
2	0.47711	0.46767	0.45841	0.44933	0.44043	0.43171	0.42316
3	0.32956	0.31982	0.31037	0.30119	0.29229	0.28365	0.27527
4	0.22764	0.21871	0.21014	0.20190	0.19398	0.18637	0.17907
5	0.15724	0.14957	0.14227	0.13534	0.12873	0.12246	0.11648
6	0.10861	0.10228	0.09633	0.09072	0.08543	0.08046	0.07577
7	0.07502	0.06995	0.06522	0.06081	0.05670	0.05287	0.04929
8	0.05182	0.04783	0.04416	0.04076	0.03763	0.03474	0.03206
9	0.03579	0.03271	0.02990	0.02732	0.02497	0.02282	0.02086
10	0.02472	0.02237	0.02024	0.01832	0.01657	0.01500	0.01357
11	0.01708	0.01530	0.01370	0.01228	0.01100	0.00985	0.00883
12	0.01180	0.01046	0.00928	0.00823	0.00730	0.00647	0.00574
13	0.00815	0.00715	0.00628	0.00552	0.00484	0.00425	0.00374
14	0.00563	0.00489	0.00425	0.00370	0.00321	0.00279	0.00243
15	0.00389	0.00335	0.00288	0.00248	0.00213	0.00184	0.00158
16	0.00269	0.00229	0.00195	0.00166	0.00142	0.00121	0.00103
17	0.00185	0.00156	0.00132	0.00111	0.00094	0.00079	0.00067
18	0.00128	0.00107	0.00089	0.00075	0.00062	0.00052	0.00044
19	0.00088	0.00073	0.00061	0.00050	0.00041	0.00034	0.00028
20	0.00061	0.00050	0.00041	0.00034	0.00027	0.00022	0.00018
21	0.00042	0.00034	0.00028	0.00022	0.00018	0.00015	0.00012
22	0.00029	0.00023	0.00019	0.00015	0.00012	0.00010	0.00008
23	0.00020	0.00016	0.00013	0.00010	0.00008	0.00006	0.00005
24	0.00014	0.00011	0.00009	0.00007	0.00005	0.00004	0.00003
25	0.00010	0.00007	0.00006	0.00005	0.00004	0.00003	0.00002
26	0.00007	0.00005	0.00004	0.00003	0.00002	0.00002	0.00001
27	0.00005	0.00004	0.00003	0.00002	0.00002	0.00001	0.00001
28	0.00003	0.00002	0.00002	0.00001	0.00001	0.00001	0.00001
29	0.00002	0.00002	0.00001	0.00001	0.00001	0.00001	0.00000
30	0.00002	0.00001	0.00001	0.00001	0.00000	0.00000	0.00000

			Continuous Interest Rates				
Year	44%	45%	46%	47%	48%	49%	50%
1	0.64404	0.63763	0.63128	0.62500	0.61878	0.61263	0.60653
2	0.41478	0.40657	0.39852	0.39063	0.38289	0.37531	0.36788
3	0.26714	0.25924	0.25158	0.24414	0.23693	0.22993	0.22313
4	0.17204	0.16530	0.15882	0.15259	0.14661	0.14086	0.13534
5	0.11080	0.10540	0.10026	0.09537	0.09072	0.08629	0.08208
6	0.07136	0.06721	0.06329	0.05961	0.05613	0.05287	0.04979
7	0.04596	0.04285	0.03996	0.03725	0.03474	0.03239	0.03020
8	0.02960	0.02732	0.02522	0.02328	0.02149	0.01984	0.01832
9	0.01906	0.01742	0.01592	0.01455	0.01330	0.01216	0.01111
10	0.01228	0.01111	0.01005	0.00910	0.00823	0.00745	0.00674
11	0.00791	0.00708	0.00635	0.00568	0.00509	0.00456	0.00409
12	0.00509	0.00452	0.00401	0.00355	0.00315	0.00279	0.00248
13	0.00328	0.00288	0.00253	0.00222	0.00195	0.00171	0.00150
14	0.00211	0.00184	0.00160	0.00139	0.00121	0.00105	0.00091
15	0.00136	0.00117	0.00101	0.00087	0.00075	0.00064	0.00055
16	0.00088	0.00075	0.00064	0.00054	0.00046	0.00039	0.00034
17	0.00056	0.00048	0.00040	0.00034	0.00029	0.00024	0.00020
18	0.00036	0.00030	0.00025	0.00021	0.00018	0.00015	0.00012
19	0.00023	0.00019	0.00016	0.00013	0.00011	0.00009	0.00007
20	0.00015	0.00012	0.00010	0.00008	0.00007	0.00006	0.00005
21	0.00010	0.00008	0.00006	0.00005	0.00004	0.00003	0.00003
22	0.00006	0.00005	0.00004	0.00003	0.00003	0.00002	0.00002
23	0.00004	0.00003	0.00003	0.00002	0.00002	0.00001	0.00001
24	0.00003	0.00002	0.00002	0.00001	0.00001	0.00001	0.00001
25	0.00002	0.00001	0.00001	0.00001	0.00001	0.00000	0.00000
26	0.00001	0.00001	0.00001	0.00000	0.00000	0.00000	0.00000
27	0.00001	0.00001	0.00000	0.00000	0.00000	0.00000	0.00000

Appendix D

DISCOUNT FACTORS FOR SUMS THAT OCCUR UNIFORMLY OVER ONE-YEAR PERIODS AFTER TIME ZERO

	Continuous Interest Rates					
Year	5%	10%	15%	20%	25%	30%
0−1	0.97541	0.95163	0.92861	0.90635	0.88480	0.86394
1−2	0.92784	0.86107	0.79927	0.74205	0.68908	0.64002
2−3	0.88259	0.77913	0.68793	0.60754	0.53666	0.47414
3−4	0.83954	0.70498	0.59211	0.49741	0.41795	0.35125
4−5	0.79860	0.63789	0.50963	0.40725	0.32550	0.26021
5−6	0.75965	0.57719	0.43865	0.33343	0.25350	0.19277
6−7	0.72260	0.52226	0.37755	0.27299	0.19742	0.14281
7−8	0.68736	0.47256	0.32496	0.22350	0.15375	0.10579
8−9	0.65384	0.42759	0.27969	0.18299	0.11974	0.07837
9−10	0.62195	0.38690	0.24073	0.14982	0.09326	0.05806
10−11	0.59162	0.35008	0.20720	0.12266	0.07263	0.04301
11−12	0.56276	0.31677	0.17834	0.10043	0.05656	0.03186
12−13	0.53532	0.28662	0.15350	0.08222	0.04405	0.02361
13−14	0.50921	0.25935	0.13212	0.06732	0.03431	0.01749
14−15	0.48438	0.23467	0.11371	0.05511	0.02672	0.01296
15−16	0.46075	0.21234	0.09788	0.04512	0.02081	0.00960
16−17	0.43828	0.19213	0.08424	0.03694	0.01621	0.00711
17−18	0.41691	0.17385	0.07251	0.03025	0.01262	0.00527
18−19	0.39657	0.15730	0.06241	0.02476	0.00983	0.00390
19−20	0.37723	0.14233	0.05372	0.02028	0.00765	0.00289
20−21	0.35883	0.12879	0.04623	0.01660	0.00596	0.00214
21−22	0.34133	0.11653	0.03979	0.01359	0.00464	0.00159
22−23	0.32469	0.10544	0.03425	0.01113	0.00362	0.00118
23−24	0.30885	0.09541	0.02948	0.00911	0.00282	0.00087
24−25	0.29379	0.08633	0.02537	0.00746	0.00219	0.00065
25−26	0.27946	0.07811	0.02184	0.00611	0.00171	0.00048
26−27	0.26583	0.07068	0.01880	0.00500	0.00133	0.00035
27−28	0.25287	0.06395	0.01618	0.00409	0.00104	0.00026
28−29	0.24053	0.05787	0.01393	0.00335	0.00081	0.00019
29−30	0.22880	0.05236	0.01199	0.00274	0.00063	0.00014
30−35	0.19742	0.03918	0.00782	0.00157	0.00032	0.00006
35−40	0.15375	0.02376	0.00369	0.00058	0.00009	0.00001
40−45	0.11974	0.01441	0.00174	0.00021	0.00003	0.00000
45−50	0.09326	0.00874	0.00082	0.00008	0.00001	0.00000
50−60	0.06460	0.00426	0.00029	0.00002	0.00000	0.00000
60−70	0.03918	0.00157	0.00006	0.00000	0.00000	0.00000
70−80	0.02376	0.00058	0.00001	0.00000	0.00000	0.00000
80−90	0.01441	0.00021	0.00000	0.00000	0.00000	0.00000
90−100	0.00874	0.00008	0.00000	0.00000	0.00000	0.00000

INDEX